Universe 26

We Got It All Wrong

Heinrich Wilson, Elias Verdan

Heinrich Wilson Publishing

All images in this book are symbolic illustrations and not original photographs of the Universe 25 experiment.

Contents

FOREWORD

In 1968, an American behavioral researcher named John B. Calhoun began what would become one of the most disturbing experiments in the history of science. He built a paradise. Not for humans, but for mice. A perfectly controlled environment where every need was met. Unlimited food, unlimited water, comfortable nesting space, no predators, no disease, no threat of any kind. He called it Universe 25. It was, by every measurable standard, a utopia.

The mice thrived. At first. They bred rapidly, formed social groups, established territories, and behaved exactly as healthy populations do when given everything they need to survive. But then something happened that Calhoun hadn't anticipated, or perhaps had anticipated all too well. The population hit a critical threshold, and the society began to collapse. Not because of scarcity. Not because of disease. Not because of any external threat. The collapse came from within.

Males became aggressive without cause, attacking others despite having no territory to defend and no resources to compete for. Females withdrew from social life, abandoning nests and refusing to raise their young. A group Calhoun called the "beautiful ones" emerged. Mice who had completely disengaged from the population, spending their days doing nothing but grooming themselves in isolated corners while their civilization crumbled around them. Reproduction slowed, then stopped. Social bonds dissolved. Purpose evaporated. And eventually, every single mouse in Universe 25 died. Not from starvation. Not from disease. From the total psychological collapse of a society that had everything it needed and no reason left to exist.

Calhoun called this phenomenon the behavioral sink. He published his findings, the scientific community debated them, and the world mostly moved on. It was, after all, just an experiment with mice.

But what if it wasn't just about mice?

This book asks a question that most people will instinctively resist, not because it's irrational, but because the implications are too uncomfortable to sit with. What if Calhoun didn't invent anything new? What if he accidentally replicated, in miniature, an experiment that was already running? One that has been running for billions of years? What if the cage isn't a metal box in a laboratory in Maryland? What if the cage is Earth? And what if we are the mice?

Before you dismiss that as metaphor or provocation, consider something that has troubled physicists, philosophers, and technologists for decades.

In 2003, the philosopher Nick Bostrom published a paper that would become one of the most discussed thought experiments of the twenty first century. His argument was statistical. If any civilization, anywhere, at any point in history, develops the technological capability to simulate conscious beings inside a constructed reality, and if they run more than one such simulation, then the number of simulated beings in existence would vastly outnumber the "real" ones. The mathematical conclusion is uncomfortable. The probability that you are living in base reality, rather than inside a simulation, is almost zero.

Bostrom's simulation hypothesis was debated, expanded, and refined by physicists, philosophers, and technologists around the world. Elon Musk publicly stated that the odds of us living in base reality are "one in billions." Neil deGrasse Tyson put the probability at fifty fifty. The physicist Sylvester James Gates discovered what appeared to be error correcting codes, the same type used in computer software, embedded in the equations that describe our universe. The deeper we look at reality, the more it resembles something that was built.

And then there are the glitches.

The Mandela Effect. The phenomenon where large groups of people share identical memories of events that apparently never happened, or remember details of the world that don't match current reality. Millions of people remember Nelson Mandela dying in prison in the 1980s. He didn't. Millions remember the Berenstain Bears spelled differently. They swear the logo on Fruit of the Loom had a cornucopia. It doesn't, and the company says it never did. These aren't cases of one confused individual. These are mass shared memories, consistent in their details, spanning cultures and continents.

If our reality is a designed system, the Mandela Effect starts to look less like faulty memory and more like evidence of revision. Patches in the code. Updates to the simulation where the new version doesn't quite match what the subjects remember from before the change. If you have ever had the unshakeable feeling that something about the world is different from how you remember it, you are not alone. And you may not be wrong.

But the simulation hypothesis, even in its most compelling form, usually stops at one question. Are we in a simulation? This book goes further. It asks the question that almost nobody asks. If this is a constructed reality, what is it for?

A simulation implies purpose. Software is written to do something. An environment is constructed for a reason. And if you look at the structure of our reality, a closed system, finite resources, a population of conscious beings placed inside it, conditions calibrated precisely to sustain life, it doesn't look like a game. It doesn't look like entertainment. It looks like an experiment.

It looks, in fact, exactly like Universe 25.

And like any experiment, the variables are not equal. Look at the world around you. Some people are born into wealth and comfort. Others are born into poverty and war. Some survive accidents that should have killed them. Others die from causes that should have been harmless. Some people seem to move through life with an invisible hand guiding them toward success, while others with equal ability and equal effort are met with nothing but resistance. We call it luck. We call it fate. We call it unfair.

But a researcher doesn't call it any of those things. A researcher calls it variable distribution. You don't give every subject in an experiment the same starting conditions, because then you can't measure anything. You place some mice near the food and others in the far corner. You give some advantages and others disadvantages. You create inequality deliberately, because inequality produces data. Every disparity in the human experience, every inexplicable stroke of fortune or misfortune, every moment where life seems impossibly unfair, makes no sense in a just universe. But it makes perfect sense in a controlled experiment.

John B. Calhoun built a sealed environment with perfect conditions and placed living creatures inside to observe what would happen. Whoever, or whatever, constructed our reality did the same thing. The scale is different. The subjects are different. The cage is a planet instead of a metal enclosure. But the structure is identical. A contained population inside a designed habitat, provided with everything needed for survival, observed from the outside.

And here is the part that should keep you awake tonight. Calhoun ran his experiment twenty five times. The mice always destroyed themselves. Every single time. He changed variables, adjusted conditions, modified the setup, and the outcome was always the same. Collapse. Extinction. The behavioral sink was not an anomaly. It was inevitable.

If our reality is an experiment, it raises the possibility that it has been run before. That previous iterations existed, played out, and ended. That what we think of as the history of life on Earth, punctuated by mass extinctions and near total annihilation events, is not random catastrophe but data. Results from previous runs. The Permian extinction, which wiped out ninety six percent of all marine species. The asteroid that ended the age of the dinosaurs. The Toba supervolcano that nearly erased humanity seventy thousand years ago. The Bronze Age Collapse. The Black Death. Each one a reset. Each one a failed run cleared from the cage before the next one began.

Calhoun's twenty fifth universe was the one where he stopped intervening. He let it run to completion. He watched every phase of the collapse without adjusting a single variable. Every mouse died.

The question this book will ask, and attempt to answer, is whether we are living inside Universe 26. The next run. The latest iteration of an experiment that has been running for far longer than human history, far longer than the existence of our species, possibly far longer than the existence of life itself on this planet. An experiment that has been reset before, that has been adjusted before, and that may now be in its final, uninterrupted phase.

Everything you think you know about progress, about civilization, about the meaning of human existence, about why we are here, this book will challenge all of it. Not with speculation. Not with conspiracy. With the convergence of behavioral science, quantum physics, historical record, religious tradition, and the observable trajectory of modern civilization, all pointing toward a single, terrifying conclusion.

We got it all wrong.

Chapter One

UNIVERSE 25

To understand where humanity might be heading, we first need to understand where a colony of mice already went. Not in the wild. Not in nature. In a carefully designed world built specifically for them by a man who wanted to answer one of the biggest questions in behavioral science. What happens to a society when every problem is solved?

John B. Calhoun was an American ethologist and behavioral researcher working at the National Institute of Mental Health. He had spent years studying the behavior of rats and mice in controlled environments, observing how populations grow, interact, and organize themselves when placed in confined spaces. But by the late 1960s, Calhoun wanted to push his research further than anyone had gone before. He didn't want to study how animals cope with hardship. He wanted to study what happens when hardship is removed entirely.

In 1968, he built what he described as a mouse utopia. A physical enclosure designed to be the perfect world for its inhabitants. It was a square metal structure, roughly four and a half feet on each side, with walls high enough that no mouse could escape. Inside, Calhoun installed everything a mouse population could ever need. There were

256 nesting boxes, each designed to house up to fifteen mice. There were endless tunnels and pathways connecting the living spaces. Food dispensers were placed throughout the environment, always full, always accessible. Water was unlimited. Temperature was kept comfortable. The enclosure was cleaned regularly to prevent the spread of disease. Predators were nonexistent. There was nothing inside Universe 25 that could kill a mouse, except another mouse.

Calhoun introduced four pairs of healthy mice into this paradise. Four males and four females, all young, all in perfect health, all carefully selected for their genetic fitness. These eight mice were the founders. The Adams and Eves of a new world.

For the first few months, the mice did what mice do. They explored their environment, established territories, and began to breed. Calhoun observed this phase carefully, noting that the mice quickly adapted to their surroundings and began forming the social structures that are natural to their species. Males claimed nesting boxes and defended small territories. Females selected mates and began producing litters. The population grew slowly but steadily, doubling roughly every fifty five days.

This was what Calhoun called the exploitation phase. The mice were discovering their world, claiming space, and reproducing at a healthy rate. Everything was functioning exactly as nature intended. The population was small enough that every mouse had access to more space, food, and resources than it could ever use. There was no competition because there was no need for competition. There was enough of everything for everyone.

By around day 315, the population had grown to approximately 620 mice. The enclosure was nowhere near full. The nesting boxes could have supported over 3,800 mice. The food supply was effectively infinite. There was no rational, physical reason for anything to go wrong. And yet, this was the moment Calhoun noticed the first signs of trouble.

The population growth began to slow. Not because mice were dying at a higher rate, but because they were breeding less. New litters were getting smaller. Some females were abandoning their young entirely, something almost unheard of in healthy mouse populations. Males who had previously maintained territories and social roles began behaving erratically. Some became hyper aggressive, attacking others without provocation, without any territorial dispute, without any apparent reason at all. Others withdrew completely, retreating to unused sections of the enclosure and refusing to interact with the rest of the population.

Calhoun had seen versions of this before in his earlier experiments with rats. But the speed and completeness of the breakdown in Universe 25 surprised even him. The mice were not running out of anything. They were not sick. They were not threatened. They were simply falling apart as a society, and they were doing it in the most perfect conditions any mouse population had ever been given.

The social structure that had formed so naturally in the early days was disintegrating. The territories that males had established were being abandoned or overrun. The hierarchies that had governed mating and nesting were collapsing. Younger mice, born into an environment where all the social roles were already filled by older, established mice, found themselves with nothing to do. No territory to claim. No role to play. No purpose to fulfill. They had been born into paradise, and paradise had nothing left to offer them.

This was the beginning of what Calhoun would later call the behavioral sink. The point at which a population, despite having every physical need met, begins to psychologically and socially self destruct. It was not caused by scarcity. It was not caused by external pressure. It was caused by something far more disturbing. The complete removal of challenge, struggle, and meaning from life.

And it was only going to get worse.

What happened next in Universe 25 followed a pattern so predictable that Calhoun would later describe it as inevitable. Once the behavioral sink took hold, there was no recovery. The society didn't stabilize. It didn't adapt. It didn't find a new balance. It simply continued to fall apart, each phase worse than the one before, until there was nothing left.

The aggression among males escalated to levels that served no biological purpose. In a normal mouse population, males fight over territory, over mates, over access to food. These conflicts have a logic to them. They establish hierarchies, determine who breeds and who doesn't, and ultimately strengthen the genetic fitness of the group. But in Universe 25, the fighting had no logic. Males attacked each other in areas where there was no territory to defend. They attacked females. They attacked juveniles. Some mice bore so many bite wounds that their fur was permanently scarred. Others were attacked so frequently that they simply stopped moving, sitting motionless in open spaces as if they had given up entirely.

At the same time, a different group of males took the opposite path. Instead of becoming aggressive, they became invisible. These mice withdrew from the population completely. They didn't fight. They didn't mate. They didn't interact with any other mouse. They moved to the least populated areas of the enclosure, often the upper levels of the nesting boxes where few other mice ventured, and spent their days in total isolation. The only activity they engaged in was grooming. They ate, they slept, and they cleaned their fur. Over and over. Their coats were immaculate compared to the battle scarred majority of the population. They were physically pristine and socially nonexistent.

Calhoun called them the beautiful ones. The name was not a compliment. It was a clinical observation with a dark edge to it. These mice looked perfect because they did nothing. They had no scars because they never engaged. They had no worn fur because they never competed, never mated, never fought, never participated in any social activity whatsoever. They were alive in the biological sense, but in every meaningful behavioral sense, they had already stopped existing. They were the living dead of Universe 25.

The females deteriorated in a different but equally devastating way. In healthy mouse populations, females are fiercely protective mothers. They build nests with care, nurse their young attentively, and defend their litters against any threat. In the early days of Universe 25, the females behaved exactly this way. But as the behavioral sink deepened, maternal instinct began to collapse.

Females started abandoning litters before the pups were fully weaned. Some stopped building proper nests entirely, giving birth in exposed, unprotected areas. Others became aggressive toward their own young, something that would be almost unthinkable in a functioning mouse society. Pup mortality skyrocketed, not because there was anything wrong with the physical environment, but because the mothers had stopped caring. The drive to nurture, to protect, to ensure the survival of the next generation, was simply gone.

The younger generation, those born during the collapse, never had a chance. They entered a world where every social role was either occupied or meaningless. There were no territories to claim because the concept of territory had broken down. There were no mates to compete for because mating had become rare and disordered. There were no elders to learn from because the older mice were either violently unstable or completely withdrawn. These young mice grew up without ever learning how to be mice. They never learned how to court a mate. They never learned how to raise young. They never learned how to establish a role within a community, because the community no longer existed in any functional sense.

This was perhaps Calhoun's most chilling finding. Even if you could somehow fix the environment, even if you could remove the aggressive mice and rehabilitate the withdrawn ones, the younger generation had never acquired the social skills necessary to rebuild. They didn't know how. The knowledge, the instinct, the behavioral programming that allows a species to function as a society had not been passed down. It was lost. And once it was lost, it could not be recovered.

The population peaked at 2,200 mice around day 560. The enclosure could have supported nearly twice that number. There was still plenty of food. There was still plenty of water. There was still plenty of nesting space. But the population never grew beyond that point. It didn't plateau and stabilize. It peaked, and then it began to die.

From day 560 onward, deaths outpaced births. The few mice that were still reproducing were producing smaller and smaller litters. Many pups didn't survive because their mothers had abandoned them. The mice that were still alive were aging, and no healthy new generation was replacing them. The beautiful ones continued to groom themselves in their isolated corners. The aggressive males continued their pointless violence. The females continued to withdraw. And the population continued to shrink.

Calhoun watched all of this without intervening. That was the point of Universe 25. Previous experiments had involved adjustments, changes to the environment, introductions of new mice, modifications to the space. Universe 25 was designed to run without interference. Calhoun wanted to see the full trajectory. He wanted to know what happens when you build a perfect world and then let nature take its course without ever stepping in.

What happens is extinction.

The last living mouse in Universe 25 died on day 1,780. It had taken less than five years for a population that had been given literally everything to destroy itself completely. Not a single mouse survived. Not a single pup was born in the final months. The world Calhoun had built with such care, the paradise he had designed to eliminate every possible source of suffering, ended in total and absolute emptiness.

When Calhoun published his findings, the scientific community paid attention. His work was discussed, debated, and cited widely. Comparisons to human society were immediate and obvious, though many researchers cautioned against drawing direct parallels between mouse behavior and human civilization. Mice are not people, they argued. Human societies are far more complex, far more adaptable, far more capable of recognizing and responding to problems before they become terminal.

Calhoun heard those arguments. He acknowledged the differences between mice and humans. But he also noted something that his critics often overlooked. He had not run this experiment once. He had run variations of it twenty five times. Twenty five different configurations, twenty five different setups, twenty five different starting conditions. And the result was always the same. Every single time, the population collapsed. Every single time, the behavioral sink took hold. Every single time, the society destroyed itself from within.

The number of the experiment was not random. It was not called Universe 25 because it sounded good. It was called Universe 25 because it was the twenty fifth attempt. Twenty five times, Calhoun had built a world, populated it, and watched it fail. Twenty five controlled experiments, all ending the same way. That is not an anomaly. That is not bad luck. That is a pattern. And patterns, in science, are the closest thing we have to certainty.

Calhoun spent the rest of his career warning anyone who would listen that human civilization was showing the early signs of the same trajectory. He pointed to rising population density in cities, to the breakdown of community structures, to the increasing isolation of individuals within increasingly crowded environments. He argued that the behavioral sink was not unique to mice. It was a universal response to a specific set of conditions, and those conditions were being replicated in human society on a global scale.

Most people didn't listen. Calhoun was respected in his field, but his warnings were treated as interesting speculation rather than urgent prophecy. The world had bigger things to worry about. The Cold War, economic growth, technological progress. The idea that humanity might be heading toward the same fate as a colony of mice in a metal box seemed absurd.

That was over fifty years ago. Since then, birth rates across the developed world have fallen to historic lows. Loneliness has been declared a public health crisis. Suicide rates have climbed steadily in the most prosperous nations on earth. Social media has created a generation of beautiful ones, disengaged from reality, obsessed with self image, participating in nothing. Political and social structures are fragmenting. Trust in institutions has collapsed. And the global population, for the first time in modern history, is approaching the point where it may begin to shrink.

Everything Calhoun warned about is happening. Everything he observed in his mice is observable in us. The only question is whether we are willing to see it.

And whether, like the mice, we are already past the point of return.

Chapter Two

THE BEHAVIORAL SINK

The term behavioral sink has become one of the most quoted phrases in the history of behavioral science, but it is also one of the most misunderstood. Most people who have heard of Calhoun's work reduce it to a simple idea. Too many mice in a small space leads to chaos. Overcrowding causes violence. Cramped conditions drive populations insane. But that is not what Calhoun found. That is not what the behavioral sink actually means. And getting it wrong matters, because the real lesson is far more disturbing than the simplified version.

Universe 25 was never overcrowded. This is the single most important detail that most people overlook when discussing the experiment. The enclosure was designed to comfortably house over 3,800 mice. The population peaked at 2,200. At no point did the mice run out of space. At no point did they run out of food or water. At no point were the physical conditions of the environment insufficient for the size of the population. The collapse did not happen because there were too many mice. It happened despite the fact that there was more than enough room for all of them.

This distinction matters because it changes everything about how we interpret the results. If the behavioral sink were simply a response to overcrowding, it would be a problem with a solution. Build a bigger cage. Give the mice more space. Problem solved. But Calhoun demonstrated that the problem was not physical. It was social. It was psychological. And it emerged not from scarcity, but from abundance.

The behavioral sink, as Calhoun defined it, was the progressive breakdown of social behavior in a population where all physical needs have been met. It was not about space. It was about purpose. When every survival need is satisfied, the behaviors that evolved to secure those needs do not simply disappear. They malfunction. They turn inward. They become destructive. The drives that once pushed mice to compete, to build, to protect, to nurture, had no outlet in a world where competition, building, protecting, and nurturing were unnecessary. And without an outlet, those drives curdled into something toxic.

Think of it this way. In the wild, a male mouse has a job. He finds territory, he defends it, he competes with other males for access to females, he helps ensure the survival of his genetic line. Every behavior he engages in has a purpose tied to survival. His aggression is functional. His social interactions are meaningful. His life has structure because the environment demands structure from him.

Now place that same mouse in Universe 25. The territory is already there, pre built and unlimited. There is nothing to defend it against, because there are no predators and food is everywhere. Competition for mates exists in theory, but with hundreds of females and abundant nesting space, the pressure that normally drives competitive behavior is almost nonexistent. The male mouse still has all of the biological drives he was born with. The aggression is still there. The territorial instinct is still there. The urge to compete is still there. But none of it has anywhere to go.

What Calhoun observed was that these drives didn't simply switch off. They misfired. Males who had no reason to fight began fighting anyway, but without the structure that normally governs conflict. In the wild, a fight between two males follows rules. There is posturing, there is escalation, there is usually a clear winner and a clear loser, and the loser retreats. The fight serves a function. It establishes hierarchy. In Universe 25, the fights had no function. They were random, brutal, and often targeted the weakest members of the population rather than rival males. The aggression had become pathological. It was no longer a tool for survival. It was a symptom of a society that had lost its reason to exist.

The females experienced a parallel collapse, though it expressed itself differently. In a functioning mouse society, the role of the female is centered around reproduction and nurturing. She selects a mate, builds a nest, bears young, and raises them until they are independent. This cycle gives her life structure and purpose in the same way that territorial competition gives the male's life structure and purpose. Remove the need for that structure, and the same thing happens. The drives don't disappear. They break down.

Females in Universe 25 began showing behaviors that were essentially unheard of in healthy populations. They became aggressive toward their own pups, attacking or neglecting newborns that they would normally have protected with their lives. Some females stopped nesting entirely, giving birth in open areas and walking away. Others became aggressive toward males, toward other females, toward any mouse that came near them, regardless of context. The nurturing instinct that defines female behavior in mouse populations had been replaced by hostility and withdrawal.

And then there were the beautiful ones. Calhoun devoted significant attention to this group because he recognized in them something that went beyond individual behavioral failure. The beautiful ones represented something new. They were not aggressive. They were not hostile. They were not struggling and failing to adapt. They had simply opted out. Completely. Permanently.

The beautiful ones were almost exclusively male, though some females exhibited similar behavior in the later stages of the collapse. They were easy to identify. Their fur was clean, sleek, and unmarked. In a population where most mice bore the scars of random violence, the beautiful ones were physically perfect. They had no bite wounds. No torn ears. No patches of missing fur. They looked healthier than any other mice in the enclosure, and that was precisely the problem.

Their appearance was pristine because they never did anything. They never fought, so they were never injured. They never competed for mates, so they were never challenged. They never defended territory, so they were never attacked. They never engaged in any social interaction of any kind, so they bore none of the physical evidence of a life lived among others. They had removed themselves from the population entirely, retreating to quiet, empty sections of the enclosure where they spent their days in a cycle of eating, sleeping, and grooming. Nothing else. No social behavior. No reproductive behavior. No exploratory behavior. Just maintenance of their own bodies in total isolation.

Calhoun noted that the beautiful ones were not sick. They were not injured. There was nothing physically wrong with them. Biologically, they were capable of mating, of socializing, of participating in the life of the colony. They simply didn't. They had never learned how, or they had learned and then forgotten, or they had decided, on some level that we might not fully understand, that participation was pointless. The world they had been born into offered them nothing to strive for, and so they strived for nothing.

What made the beautiful ones particularly significant to Calhoun was their irreversibility. In later phases of the experiment, when the population had declined significantly and there was, in theory, plenty of social space for new relationships and new roles to form, the beautiful ones did not re engage. They did not suddenly start mating or competing or participating. The withdrawal was permanent. Whatever psychological switch had been flipped could not be flipped back. They remained in their isolated corners, grooming themselves, until they died.

This was the detail that haunted Calhoun more than any other. It suggested that the behavioral sink was not just a temporary crisis that a population could recover from. It was a terminal condition. Once a critical mass of individuals had disengaged from society, the society could not rebuild itself even if conditions improved. The knowledge of how to function socially, how to form bonds, how to raise young, how to participate in a community, had been lost. And in a species that relies on learned social behavior passed down through generations, that loss was permanent.

Calhoun called this phase the death of societal function. The population was still alive in a biological sense. Mice were still eating, still breathing, still physically present in the enclosure. But the society was dead. The web of relationships and roles and behaviors that make a group of individuals into a functioning community had dissolved beyond repair. What remained was a collection of isolated bodies occupying the same physical space without any meaningful connection to one another.

The final phase of Universe 25 was not dramatic. There was no sudden catastrophe, no mass die off, no single event that ended the colony. It was a slow, quiet disappearance. The old mice aged and died. The few remaining young mice, raised without any functional social framework, never learned how to breed successfully. Litters became rare, then nonexistent. The population shrank month by month, year by year, until the enclosure that had once held over two thousand mice held only a handful of aging, solitary individuals who had never known what a functioning society looked like.

The last mouse died alone.

Calhoun spent years analyzing his data and trying to communicate the significance of what he had observed. He gave lectures. He published papers. He spoke to anyone who would listen about what he believed was a universal pattern, not limited to mice, not limited to any particular species, but a fundamental consequence of what happens when a population is given everything and asked to do nothing.

His most provocative argument was this. The behavioral sink was not caused by the density of the population. It was caused by the loss of functional roles within that population. When every mouse had access to unlimited resources, the behaviors that had evolved to secure those resources became meaningless. And when behavior becomes meaningless, the individual becomes meaningless. And when enough individuals lose their sense of meaning, the society collapses regardless of how perfect the physical conditions are.

He argued that the key variable was not space, not food, not physical comfort, but purpose. The mice in Universe 25 died because they had no reason to live. Not in the romantic, philosophical sense. In the behavioral sense. Their daily existence required nothing of them. They did not need to search for food. They did not need to avoid predators. They did not need to build shelter. They did not need to do anything at all. And a living creature that does not need to do anything will, eventually, stop doing everything.

Including living.

Calhoun saw this as a warning for humanity. He was careful to acknowledge the differences between mice and humans. We are more complex. We have language, culture, abstract thought, the ability to recognize problems and consciously adapt to them. But he also pointed out that complexity does not equal immunity. A more complex society that loses its sense of purpose will collapse in a more complex way, but it will still collapse. The behavioral sink does not care how intelligent the species is. It cares only whether the species still has a reason to function.

When Calhoun looked at the world around him in the 1970s and 1980s, he saw the early stages of exactly what he had observed in his mice. Growing urban populations, increasing social isolation despite increasing physical proximity, a rising generation that seemed disconnected from the roles and structures that had given previous generations their sense of meaning. He warned that if humanity continued on this trajectory, the result would be the same. Not immediately. Not in a decade. But inevitably.

He called it the first death. The death of the spirit within a society that is still physically alive. The second death, the physical extinction, was simply a matter of time after that.

Most people found the comparison alarmist. Some found it offensive. The idea that human beings, with all of our technology and intellect and creativity, could be reduced to the same behavioral patterns as mice in a metal box was too much for many to accept.

But Calhoun's data didn't care about comfort. His results didn't change depending on who was listening. Twenty five experiments.

Twenty five collapses. Twenty five extinctions. The pattern was there for anyone willing to look at it.

The question was never whether the behavioral sink could happen to humans. The question was whether it already had.

Chapter Three

THE CAGE

Every experiment needs a controlled environment. A space that is sealed off from outside interference, where variables can be measured and outcomes can be observed without contamination from external forces. Calhoun understood this. That is why he built Universe 25 the way he did. A closed system. Walls high enough that nothing could get in and nothing could get out. Everything the subjects needed to survive was provided within the enclosure. The mice never had to look beyond the walls of their world, because there was nothing they needed that wasn't already inside it.

Now look at where you are.

You are sitting on a planet that is, by any reasonable definition, a sealed enclosure. Earth is surrounded by a vacuum that no living creature can survive in without artificial life support. The atmosphere forms a thin, fragile barrier between the surface and the void of space. Beyond that barrier, there is nothing. No air. No water. No food. No warmth. Just an emptiness so vast and so hostile that even with all of our technology, we have barely managed to send a handful of humans a short distance beyond it, and even then, only inside carefully engineered capsules that recreate the conditions of the enclosure they left behind.

We cannot leave. That is the simple truth that most of us never think about, because we never need to. There is nowhere to go. The nearest planet that might, under the most optimistic projections, support some form of human habitation is Mars, and it is a frozen, airless desert with no breathable atmosphere, no liquid water on the surface, and radiation levels that would kill an unprotected human in a matter of months. Every other body in our solar system is even less hospitable. And beyond our solar system, the distances involved are so enormous that reaching even the nearest star with current technology would take tens of thousands of years.

We are, for all practical purposes, confined to this planet. The walls of our cage are not made of metal. They are made of vacuum and radiation and distances so large that the human mind cannot meaningfully comprehend them. But the effect is the same. We are inside, and we cannot get out.

And just like Calhoun's enclosure, everything we need to survive is provided within the walls.

The parallels are so precise that they would be difficult to believe if they weren't simply factual. Calhoun's mice had unlimited food. We live on a planet where the soil, the oceans, and the atmosphere combine to produce an effectively endless food supply. Humanity has never lacked the ability to feed itself. Even today, with eight billion people on the planet, global food production exceeds global food demand. The problem has never been supply. It has been distribution, politics, greed, and waste. The food is there. It has always been there.

Calhoun's mice had unlimited water. Earth's surface is roughly seventy percent water. The hydrological cycle, driven by solar energy, continuously purifies and redistributes water across the planet through evaporation and rainfall. Fresh water exists in rivers, lakes, underground aquifers, and glaciers in quantities that dwarf what humanity consumes. Again, the shortages that exist are not failures of supply. They are failures of access, infrastructure, and management. The water is there.

Calhoun's mice had a temperature controlled environment. Earth sits in what astronomers call the habitable zone, the narrow band of distance from the Sun where temperatures allow liquid water to exist on the surface. Too close, and the oceans would boil. Too far, and they would freeze solid. Our planet happens to sit in exactly the right spot. The atmosphere regulates temperature through the greenhouse effect, distributing heat from the equator toward the poles, moderating extremes, and maintaining a range that supports life across the majority of the planet's surface. The thermostat is built in.

Calhoun's mice were free from predators. Humans have effectively eliminated every natural predator that once threatened us. There is no animal on Earth that poses an existential threat to the human species. We sit at the top of every food chain on the planet. We have no natural enemies in the biological sense. The things that threaten us, disease, natural disasters, climate events, are features of the environment itself, not external predators trying to get into the cage. We are alone at the top, just as Calhoun's mice were alone inside their walls.

Calhoun's mice were free from disease, at least initially. Humanity has not achieved this completely, but we have come closer than any species in the history of life on Earth. Vaccines have eradicated or controlled diseases that once killed millions. Antibiotics have turned previously fatal infections into minor inconveniences. Surgical techniques, diagnostic technology, and pharmaceutical development have pushed human life expectancy to levels that would have been unimaginable even two hundred years ago. We have not eliminated disease, but we have reduced it to a degree that Calhoun's mice, in their sterile, controlled enclosure, would recognize as familiar.

The resemblance between Earth and a laboratory enclosure does not stop at the basic provisions. It extends to the structure of the environment itself, in ways that become increasingly difficult to dismiss as coincidence the closer you look.

Consider the design of Universe 25. Calhoun did not simply throw mice into an empty box. He built an environment with features. Nesting areas, pathways, multiple levels, different zones that the mice could explore and inhabit. The enclosure had variety. It had complexity. It gave the mice the impression of a rich, detailed world to live in. But it was still a box. No matter how many tunnels they explored or corners they discovered, the walls were always there. The mice could spend their entire lives navigating the internal complexity of their environment without ever reaching the edge of it, and most of them did. But the edge was always there, and beyond it, nothing.

Earth works the same way. We live on a planet of extraordinary variety. Mountains, oceans, deserts, forests, plains, tundra, jungles, islands, river systems, coral reefs. The sheer diversity of environments on Earth gives us the impression of a world so vast and so complex that it could never be fully explored. And for most of human history, that impression was accurate. There were always new lands to discover, new territories to map, new frontiers to push toward.

But we have now mapped every inch of it. Satellites photograph the entire surface of the planet in real time. There are no unknown continents. There are no undiscovered countries. The frontier is closed. We know exactly how big our enclosure is, and we know there is nowhere left to go within it that we have not already been. The internal complexity is extraordinary, but the boundaries are fixed. We have explored every tunnel, every corner, every level of our cage, and we have reached the walls.

And what lies beyond the walls? The same thing that lay beyond the walls of Universe 25. Nothing that can sustain us. The mice could not survive outside their enclosure. We cannot survive outside ours. The vacuum of space is our version of the laboratory floor on the other side of the metal walls. It is there, we know it exists, but it is not for us. Not yet. Perhaps not ever.

This creates a psychological condition that Calhoun would have recognized immediately. A population confined to a fixed space, no matter how large or how well provisioned that space may be, eventually reaches a point where the environment feels full. Not in the physical sense. Universe 25 was never physically full. Earth, by most meaningful metrics, is not physically full either. But there is a difference between physical capacity and psychological capacity. When every space has been claimed, when every role has been filled, when every frontier has been explored and mapped and settled, the environment begins to feel complete. Finished. Closed. And a closed environment, no matter how comfortable, eventually begins to suffocate the population inside it.

This is what Calhoun meant when he talked about the loss of functional roles. It was not just about having a job or a territory. It was about the relationship between a living creature and its environment. In the wild, that relationship is dynamic. The environment presents challenges, and the creature responds. The creature's behavior shapes the environment, and the environment shapes the creature's behavior. There is a constant back and forth, a tension, a reason to act. In a closed, perfected environment, that tension disappears. The environment asks nothing of the creature, and the creature has nothing to ask of the environment. The relationship becomes static. And static relationships, whether between individuals or between a species and its habitat, eventually die.

Look at the trajectory of human civilization through this lens. For hundreds of thousands of years, the environment demanded everything from us. We hunted, gathered, migrated, fought predators, endured ice ages, crossed oceans, built shelters from raw materials, and died in enormous numbers from threats that were constant and real. Every day was a negotiation with the environment. Stay alive. Find food. Protect your family. Survive until tomorrow. The relationship between humanity and Earth was dynamic, brutal, and meaningful.

Then we began to win. Agriculture removed the need to hunt and gather. Architecture removed the need to seek natural shelter. Medicine removed the threat of most diseases. Technology removed the need for physical labor. Each advancement was a victory, a genuine improvement in the quality of human life. Nobody would argue that these were bad things in isolation. But each one also removed a demand that the environment had placed on us. Each one reduced the tension between the species and its habitat. Each one brought us closer to the condition that Calhoun created artificially in Universe 25. A world where everything is provided and nothing is required.

We are not there yet. Not entirely. There are still billions of people on this planet who struggle daily for survival, who face real scarcity, real danger, real environmental pressure. But the trajectory is clear, and in the most developed nations on Earth, the destination is already visible. Societies where food is abundant, shelter is guaranteed, physical safety is the norm, and life expectancy stretches into the eighties and beyond. Societies where the basic challenges of survival have been solved so completely that entire generations have never experienced genuine hardship.

And it is precisely those societies, the most advanced, the most comfortable, the most perfected, where the cracks are showing first.

If the behavioral sink were simply a product of poverty, war, and suffering, you would expect the most prosperous nations to be the healthiest in every sense. The opposite is true. The countries with the highest standards of living on Earth are the ones recording the steepest declines in mental health. Anxiety and depression rates in Western Europe, North America, Japan, South Korea, and Australia have climbed steadily for decades, and the sharpest increases are among young people. The generation that has grown up with the most comfort, the most safety, and the most material abundance in human history is also the generation reporting the highest rates of psychological distress ever recorded.

Suicide rates tell the same story. Globally, suicide is now one of the leading causes of death among young people aged fifteen to twenty nine. But the distribution is not what you would expect if suffering were the primary driver. Some of the highest suicide rates in the world are found not in war zones or famine regions but in wealthy, stable, peaceful countries. South Korea, one of the most technologically advanced nations on Earth, has one of the highest suicide rates in the developed world. Japan, a country with universal healthcare, virtually zero violent crime, and one of the longest life expectancies on the planet, has struggled with epidemic levels of suicide for decades. Scandinavia, routinely cited as home to the happiest countries in the world based on material metrics, consistently reports suicide rates that are higher than the global average.

These are the perfect corners of the cage. These are the sections of the enclosure where the food is closest, the temperature is most comfortable, and the nesting conditions are the best. And these are the sections where the mice are falling apart first.

The pattern extends beyond mental health. Birth rates in the most developed nations have fallen below replacement level and continue to drop. Social isolation is rising fastest in the countries with the most connectivity. Loneliness, which researchers now recognize as a health risk comparable to smoking, is most prevalent in the societies with the most advanced communication technology. People who can reach anyone on Earth in seconds report feeling more alone than any previous generation.

Drug and alcohol addiction follow the same pattern. The opioid crisis did not begin in the poorest parts of America. It began in the comfortable ones. Suburbs, small towns, middle class communities where people had homes, jobs, and stability. The same is true across Europe, where addiction rates have climbed steadily in countries that have largely eliminated poverty, homelessness, and violent crime as widespread social problems.

Calhoun would not have been surprised by any of this. He would have recognized it immediately. The mice in Universe 25 did not begin their collapse in the worst parts of the enclosure. The collapse began in the most comfortable areas, among the mice with the easiest access to food, the most secure nesting sites, and the least competition. The mice that had it best fell apart first, because they were the first to experience the full weight of having nothing left to strive for.

The cage does not need to be cruel to destroy its inhabitants. It does not need to starve them or freeze them or expose them to danger. It just needs to be complete. It just needs to provide everything, ask nothing, and wait.

Calhoun built a cage and filled it with everything his mice could need. He watched them thrive, and then he watched them destroy themselves. He did not cause the destruction. He simply created the conditions for it and stepped back.

Whoever or whatever placed eight billion humans on a planet with breathable air, drinkable water, fertile soil, and a climate perfectly calibrated to support life did exactly the same thing. The cage is bigger. The tunnels are more complex. The nesting areas are more varied. But the walls are just as real, the provisions are just as complete, and the question is just as urgent.

What happens to the mice when the cage is perfect and there is nowhere left to go?

We already know the answer. Calhoun showed it to us twenty five times.

The only question is whether we are watching it happen again, from the inside.

Chapter Four

THE GREAT RESETS

If you accept, even as a thought experiment, the possibility that life on Earth exists inside a controlled environment, then a natural question follows. How long has the experiment been running? And has it always been the same experiment, or have there been previous versions?

The history of life on this planet provides an uncomfortable answer. Life on Earth is roughly 3.8 billion years old. In that time, it has been nearly wiped out on multiple occasions. Not gradually. Not through slow decline. Through sudden, catastrophic events that killed the vast majority of living species in geologically brief windows of time. Scientists call these mass extinction events, and the geological record documents at least five major ones, each of which fundamentally reset the trajectory of life on the planet.

These are not small disruptions. These are events that eliminated between seventy and ninety six percent of all species alive at the time. After each one, life rebuilt from whatever survived, evolving in new directions, producing new dominant species, filling the ecological gaps left by the dead. And then, eventually, another extinction event occurred, and the process started over.

If you are running a long term experiment and the subjects fail to produce the outcome you are looking for, you have a few options. You can end the experiment entirely. You can adjust the variables and try again with the same subjects. Or you can clear the environment, introduce new variables, and start a fresh run. The history of mass extinctions on Earth looks remarkably like the third option, repeated five times over hundreds of millions of years.

The first major extinction occurred roughly 445 million years ago, at the end of the Ordovician period. At that time, most complex life existed in the oceans. The dominant creatures were marine invertebrates, trilobites, brachiopods, bryozoans, and early coral systems. Then, over a span of roughly one to two million years, approximately eighty five percent of all marine species disappeared. The cause is believed to have been a combination of severe glaciation and dramatic changes in sea level and ocean chemistry. The planet cooled rapidly, ice sheets expanded, and the conditions that had supported a thriving marine ecosystem simply ceased to exist.

The second extinction came roughly seventy million years later, near the end of the Devonian period, around 375 million years ago. This one was slower, unfolding in a series of pulses over several million years, but the result was the same. Approximately seventy five percent of all species were eliminated. The Devonian had been an age of fish and the first land plants, a period of rapid diversification and expansion into new habitats. By the end of it, most of that diversity was gone. The causes remain debated, but the leading theories point to massive volcanic activity, changes in ocean oxygen levels, and the rapid spread of land plants altering atmospheric chemistry in ways that destabilized the global climate.

The third and most devastating extinction occurred 252 million years ago, at the boundary between the Permian and Triassic periods. This is sometimes called the Great Dying, and the name is not an exaggeration. An estimated ninety six percent of all marine species and seventy percent of all terrestrial vertebrate species were wiped out. Nearly everything died. The cause was almost certainly massive volcanic eruptions in what is now Siberia, events so enormous that they fundamentally altered the composition of the atmosphere and the chemistry of the oceans on a global scale. The Permian extinction came closer to ending life on Earth entirely than any event before or since. The planet very nearly became sterile.

The fourth extinction closed the Triassic period, roughly 201 million years ago, eliminating around eighty percent of all species. This event cleared the way for the dinosaurs, which had existed as relatively minor creatures during the Triassic, to become the dominant land animals on Earth. The causes are again linked to massive volcanism and associated climate disruption.

And then there is the fifth and most famous extinction, the one that ended the Cretaceous period sixty six million years ago. An asteroid roughly ten kilometers wide struck what is now the Yucatan Peninsula in Mexico, releasing energy equivalent to billions of nuclear weapons and triggering a global catastrophe that killed approximately seventy six percent of all species, including every non avian dinosaur on the planet. The dominant life forms that had ruled Earth for over 160 million years were erased in a geological instant, and the ecological vacuum they left behind was eventually filled by the mammals that would, tens of millions of years later, produce us.

Five mass extinctions over roughly four hundred million years. Five times the majority of life on Earth was destroyed. Five times the survivors rebuilt, evolved, diversified, and eventually produced new dominant species. And five times, those dominant species were wiped out and replaced.

Look at this pattern from the perspective of experimental design. Each extinction event eliminated the existing dominant life forms and created the conditions for new ones to emerge. The Ordovician extinction cleared the oceans of their first complex ecosystems, allowing new marine life to evolve in the empty space. The Devonian extinction ended the age of the great fish and the earliest land colonizers, making way for the amphibians and early reptiles. The Permian extinction, the most severe of all, killed nearly everything and reset the board so completely that entirely new categories of life emerged in the aftermath. The Triassic extinction removed the competitors that had been holding the dinosaurs in check, allowing them to dominate the planet for the next 160 million years. And the Cretaceous extinction removed the dinosaurs, clearing the way for small, adaptable mammals to inherit the Earth.

Each reset produced a more complex set of subjects than the one before. Simple marine invertebrates gave way to fish. Fish gave way to amphibians and reptiles. Reptiles gave way to dinosaurs. Dinosaurs gave way to mammals. And mammals eventually produced a species capable of language, abstract thought, technology, and the ability to ask questions about its own existence. Each iteration of life on Earth has been more sophisticated, more behaviorally complex, and more aware of its environment than the last.

If this is an experiment, the experimenter has been refining the subjects for nearly four billion years.

But the major extinctions are only part of the story. Between the great resets, there have been smaller disruptions, near misses, and localized catastrophes that thinned populations and altered evolutionary trajectories without triggering a full planetary extinction. The fossil record is littered with these smaller events, periods of elevated extinction that don't rise to the level of the big five but still killed millions of species and reshaped the course of life on Earth. These were not full resets. They were adjustments. Corrections to the experiment while it was still running.

And then humanity arrived. And the resets did not stop. They simply changed scale.

The Toba supervolcano erupted roughly 74,000 years ago in what is now Indonesia. It was one of the largest volcanic events in the last two million years, ejecting enormous quantities of ash and sulfur into the atmosphere and triggering a volcanic winter that may have lasted a decade or more. Global temperatures plummeted. Plant life died on a massive scale. Food sources collapsed. And the human population, already small and spread thinly across parts of Africa and possibly Asia, was pushed to the edge of extinction.

Genetic studies suggest that the human population may have been reduced to as few as three to ten thousand breeding individuals in the aftermath of Toba. Some researchers put the number even lower. The entire species, every human being alive on the planet, could have fit inside a single modern sports stadium. We came that close to vanishing entirely.

But we didn't. A small number survived, and from that tiny population, the entire human species rebuilt. Every person alive today is descended from that handful of survivors. The experiment nearly ended, but it didn't. The cage was almost cleared, but a remnant was left. Just enough to start again.

Toba was not the last time humanity was nearly erased. The Younger Dryas, a sudden and severe cold period roughly 12,800 years ago, struck just as early human civilizations were beginning to develop agriculture and permanent settlements. Temperatures dropped dramatically, particularly in the Northern Hemisphere, and remained cold for over a thousand years. The cause is still debated. Some scientists point to a massive influx of freshwater into the North Atlantic disrupting ocean circulation. Others suggest a comet impact or airburst event. Whatever the cause, the result was devastating. Emerging human societies were thrown back into survival mode. Populations declined. Progress stalled. The first steps toward civilization were interrupted and delayed by over a millennium.

Then there was the Bronze Age Collapse, around 1200 BC. This was not a natural disaster. It was the sudden and near simultaneous destruction of virtually every major civilization in the eastern Mediterranean and Near East. The Mycenaeans, the Hittites, the Egyptian New Kingdom, the Kassite dynasty of Babylon, the cities of Ugarit and Alalakh, all of them either collapsed entirely or were severely diminished within a span of roughly fifty years. Literacy was lost in several regions. Trade networks that had connected the ancient world for centuries disintegrated. Cities that had stood for generations were abandoned or burned.

The causes of the Bronze Age Collapse are still debated by historians. Invasion by the mysterious Sea Peoples, drought, famine, earthquakes, internal rebellion, disruption of trade routes, some combination of all of these. What is not debated is the result. In the span of a single lifetime, the most advanced human civilizations on the planet were reduced to rubble. Centuries of knowledge, culture, and technological progress were lost. Humanity did not go extinct, but it was set back dramatically. The world that emerged from the collapse was simpler, poorer, and less connected. It took centuries to rebuild what had been destroyed in decades.

The Black Death arrived in Europe in 1347 and killed an estimated seventy five to two hundred million people over the next several years. At its peak, it eliminated roughly a third of the entire European population. Some regions lost far more. Entire villages were emptied. Cities lost half their inhabitants. The social, economic, and political structures of medieval Europe were shattered. Labor shortages transformed feudal economies. Religious authority was questioned as prayers failed to stop the plague. The psychological trauma of watching a third of the known world die in agony within a few years reshaped European culture in ways that are still visible today.

And these are just the most well known examples. The Plague of Justinian in the sixth century killed an estimated twenty five to fifty million people across the Byzantine Empire and beyond. The arrival of Europeans in the Americas triggered the death of an estimated ninety percent of the indigenous population through disease, violence, and displacement. The Spanish Flu of 1918 killed between fifty and one hundred million people in less than two years.

Over and over, human populations have been reduced, shattered, and rebuilt. The details change. The mechanisms vary. But the pattern is consistent. A population grows, develops complexity, builds social structures, and then is dramatically reduced by an event that seems to come from nowhere. The survivors start again, often from almost nothing, and the cycle repeats.

In Calhoun's earlier experiments, he intervened. He adjusted variables, introduced new mice, modified the environment, and reset the conditions to see if a different starting point would produce a different result. It never did, but he kept trying. Each universe before Universe 25 involved some level of interference from the experimenter. Some adjustment. Some attempt to correct the trajectory before it reached its conclusion.

The history of near extinctions on Earth looks the same. Repeated interventions. Populations reduced but not eliminated. Conditions altered but the experiment allowed to continue. As if someone were watching, taking notes, and deciding when to adjust and when to let things play out. As if each catastrophe were not random bad luck but a deliberate recalibration. A new variable introduced. A failed approach discarded. A fresh attempt with the surviving population.

Calhoun ran twenty five versions of his experiment. He adjusted, modified, and reset twenty four times before finally deciding, with Universe 25, to stop interfering and let the population run to completion.

The question that this pattern raises is whether the same decision has been made with us. Whether the resets that have punctuated human history, the volcanic winters and plagues and civilizational collapses, were the equivalent of Calhoun's earlier interventions. Attempts to correct the experiment. Efforts to push the population in a different direction. And whether, at some point, the experimenter decided, as Calhoun did, to stop.

To let this run play out to its end.

Chapter Five

THE NEW MICE

Every version of Calhoun's experiment began the same way. A small number of carefully selected subjects were introduced into the enclosure and left to establish themselves. The mice were healthy, genetically fit, and capable of reproducing. They were given time to explore their environment, find their place within it, and begin building the social structures that would define their colony. The early days were always the same. Curiosity, exploration, adaptation, growth.

Humanity's story begins the same way.

We do not know exactly when the first anatomically modern humans appeared, but the fossil record places the emergence of Homo sapiens in Africa somewhere between 200,000 and 300,000 years ago. For the vast majority of that time, humans were a small, scattered population of hunter gatherers, numbering in the thousands or tens of thousands at most. We were not the dominant species on the planet. We were not even the only species of human. Neanderthals occupied Europe and parts of Asia. Denisovans lived across a wide range of eastern Eurasia. Homo erectus may have still survived in isolated pockets of Southeast Asia. We shared the cage with other versions of ourselves, other experimental subjects running on slightly different biological programming.

And then, one by one, they all disappeared.

The Neanderthals, who had thrived in Europe for hundreds of thousands of years, vanished roughly 40,000 years ago. The Denisovans disappeared around the same period, though the timeline is less certain. Homo erectus, the longest surviving human species other than our own, eventually died out as well. The reasons for their extinctions are debated. Climate change, competition with Homo sapiens, interbreeding and absorption, disease, or some combination of all of these. But the result is clear. Of all the human species that once existed on this planet, only one remains. Us.

In experimental terms, the other subjects were eliminated. Whether they failed on their own or were outcompeted by a more successful variant, the outcome is the same. The experiment narrowed its focus to a single species. Every other version of the human model was discarded, and all resources, all environmental potential, all future development was concentrated on Homo sapiens.

We are the surviving strain. The version that made it through the selection process.

From that small surviving population, humanity expanded in a pattern that Calhoun would have found instantly familiar. We spread across the available environment, slowly at first, then with increasing speed. We left Africa and moved into the Middle East, then into Europe and Asia, then into Australia, and eventually across the land bridge into the Americas. Each new territory we entered was explored, settled, and incorporated into the growing range of the species. We were in the exploitation phase, exactly as Calhoun's mice were in their first months inside Universe 25. Discovering the environment, claiming space, growing the population, establishing the social structures that would define our civilizations.

For tens of thousands of years, this phase continued. Human populations grew slowly, limited by the same pressures that limit any species in the wild. Disease, predation, famine, climate, conflict. We lived in small groups, rarely more than a few hundred individuals, and our survival was never guaranteed. Every generation faced the real possibility of being the last. The environment demanded constant effort, constant adaptation, constant vigilance. And in return, it gave us something that Calhoun's mice also had in their early days. Purpose. Direction. A reason to wake up, to act, to struggle, to survive.

The parallels between early humanity and the early phase of Universe 25 are striking in their precision. Calhoun's founding mice explored every corner of the enclosure before settling into territories. Early humans did the same with the planet, spending tens of thousands of years migrating into every habitable region on Earth before settling into permanent communities. The mice formed social hierarchies based on strength, dominance, and access to mates. Early human societies did exactly the same, organizing around tribal leaders, warriors, and those who controlled access to food, shelter, and reproductive partners. The mice established territories and defended them against rivals. Humans built villages, then cities, then nations, and fought wars over every inch of ground they claimed.

Even the timeline follows a similar curve. Calhoun's mice grew slowly in the early phase, taking months to fill even a fraction of the available space. Then growth accelerated as the population reached a critical mass and social structures stabilized. Humanity followed the same trajectory on a vastly larger timescale. For most of our existence, the global population grew so slowly that it took until roughly 10,000 BC to reach an estimated five million people. Then agriculture was invented, and growth began to accelerate. By the time of the Roman Empire, the global population had reached somewhere around 200 to 300 million. By 1800, it had reached one billion. Then the industrial revolution happened, and the curve went nearly vertical. Two billion by 1930. Three billion by 1960. Four billion by 1975. Eight billion today.

Calhoun's mice hit their population peak around day 560. The growth stopped not because the enclosure was full, but because the social fabric had begun to tear. The question hanging over humanity is whether we are approaching our own day 560, the point where growth slows not because of physical limits but because of something breaking inside the species itself. The data on birth rates, mental health, and social cohesion in the most developed nations suggests we may have already passed it.

But there is something about humanity that sets us apart from every other species that has ever lived in this cage. Something that no previous run of the experiment ever produced. Something that, depending on how you interpret it, either makes us the most advanced subjects the experiment has ever seen or the most dangerous.

We are aware that the cage exists.

No other species in the history of life on Earth has looked up and understood that there is something beyond the boundaries of its environment. Fish do not contemplate the ocean. Birds do not question the sky. Calhoun's mice never once looked at the walls of their enclosure and wondered what was on the other side, or who had built it, or why they were inside it. They lived entirely within the reality that was presented to them, without any capacity to question its nature or its purpose.

Humans do something entirely different. We have looked at the boundaries of our cage and asked questions about them since the beginning of recorded thought. We have studied the stars, mapped the cosmos, theorized about the origins of the universe, and developed mathematical frameworks to describe the laws that govern our reality. We have built telescopes that can see billions of light years into space and particle accelerators that can probe the smallest building blocks of matter. We have asked, repeatedly and with increasing sophistication, what this place is, how it works, and why we are here.

And perhaps most significantly, we are the only species that has ever conceived of the idea that something might be watching.

Every human civilization, without exception, has developed the concept of a higher power. A creator. An observer. A being or force that exists beyond the boundaries of the visible world and watches, judges, or guides the inhabitants within it. This is not a cultural quirk or a regional superstition. It is a universal feature of the human species. From the earliest cave paintings to the most sophisticated modern theologies, humans have looked at the world around them and concluded, independently and repeatedly across every continent and every era, that someone or something built this place and is paying attention to what happens inside it.

No other species has ever done this. Not even close. And while there are many explanations for why humans developed religious and spiritual beliefs, ranging from evolutionary psychology to social cohesion to simple fear of death, there is one explanation that is rarely considered because of how uncomfortable it is.

What if we are right?

What if the reason every human civilization has independently concluded that a creator is watching is because a creator is watching? Not in the supernatural sense. Not a bearded figure in the clouds. But in the same way that Calhoun watched his mice. A researcher. An observer. An intelligence on the other side of the walls, monitoring the experiment, taking notes, watching to see what the subjects do.

If Calhoun's mice had possessed human level consciousness, they would have done exactly what we have done. They would have looked at the walls of their enclosure and wondered what was beyond them. They would have noticed that food appeared without explanation, that the temperature was always comfortable, that their world seemed designed for their benefit. They would have told stories about a powerful being who created their world and provided for them. They would have developed rituals and rules and belief systems centered around this unseen provider. They would have called it God.

We will return to this idea later in this book, because it has implications that go far deeper than most people are willing to consider. For now, it is enough to note that humanity is unique among all the subjects that have ever existed inside this cage. We are the first to suspect that we are inside one.

Whether that awareness changes the outcome of the experiment, or whether it simply means we will watch ourselves collapse with our eyes open, remains to be seen.

Awareness of the cage is one thing. Understanding what it means is another. Humanity has spent thousands of years developing increasingly sophisticated ways to describe the world we live in, and with each new discovery, the picture that emerges looks less like a random accident and more like something that was put together with extraordinary precision.

Start with the basic conditions required for life to exist. Physicists have identified a set of fundamental constants that govern the behavior of the universe. The strength of gravity. The speed of light. The mass of an electron. The ratio between the electromagnetic force and the strong nuclear force. These are not flexible numbers. They are fixed values that determine how matter behaves, how atoms form, how stars burn, and how chemistry works. And they are fine tuned to a degree that is almost impossible to comprehend.

If the gravitational constant were slightly stronger, stars would burn too hot and too fast, consuming their fuel before planets could form or life could develop. If it were slightly weaker, stars would never ignite at all, and the universe would be a cold, dark expanse of drifting gas. If the strong nuclear force were fractionally different, atoms heavier than hydrogen could not exist, which means no carbon, no oxygen, no water, no life. If the electromagnetic force were slightly stronger or slightly weaker, chemical bonds would not form in the way necessary to produce the complex molecules that make biology possible.

The margin for error in these constants is not small. It is essentially zero. The probability of a universe randomly producing the exact set of values necessary to support complex life has been calculated by physicists at figures so small that they are functionally indistinguishable from impossible. And yet here we are, in a universe where every single one of these values falls within the impossibly narrow range required for us to exist.

This is known as the fine tuning problem, and it has troubled physicists for decades. There are several proposed explanations. Some scientists argue for the multiverse theory, the idea that an infinite number of universes exist with every possible combination of physical constants, and we simply happen to live in the one where the numbers work. Others suggest that the constants are not truly independent but are connected by some deeper physical law that we have not yet discovered. And others, a growing number, have pointed out that the simplest explanation is also the most uncomfortable one. The constants are fine tuned because someone tuned them.

This is not a fringe position. It is a conclusion that serious physicists have been willing to state publicly. The universe looks designed. That does not prove that it was designed, but it does mean that the most straightforward reading of the evidence points toward intention rather than accident.

Now combine this with the simulation hypothesis discussed in the foreword. If the mathematical probability of our universe existing by chance is effectively zero, and if the structure of reality at the quantum level contains features that resemble programmed systems, and if the conditions on our planet are calibrated with extraordinary precision to support life, then the idea that we are inside a constructed environment is not wild speculation. It is arguably the most rational interpretation of the available evidence.

Calhoun's mice lived in a world where the temperature was always right, where food was always available, where the physical conditions for survival were always maintained. They did not know that these conditions were being maintained by someone. They had no concept of a laboratory, a scientist, or an experiment. They simply existed in an environment that happened to be perfect for them, and they never questioned why.

We exist in an environment that happens to be perfect for us. The temperature range on Earth's surface supports liquid water and biological chemistry. The atmosphere contains the exact mixture of gases needed for respiration and radiation shielding. The magnetic field deflects solar radiation that would otherwise strip the atmosphere away. The moon stabilizes the planet's axial tilt, preventing the kind of extreme climate oscillations that would make complex life impossible. The planet sits at precisely the right distance from a stable, long lived star in a quiet region of a spiral galaxy, far from the radiation hazards of the galactic center.

Every one of these conditions is necessary for us to be here. And every one of them is in place. Not approximately. Not roughly. Precisely.

The mice never asked who built the enclosure or why the food kept appearing. But we have asked. We have been asking for as long as we have been capable of forming the question. And the answers we have come up with, across every culture and every civilization, follow a remarkably consistent pattern. Someone built this. Someone is watching. Someone has rules for how we should behave.

We will explore that pattern in depth later in this book. But the point for this chapter is simpler. Humanity is not just another species occupying the cage. We are a fundamentally different kind of subject. We are the first to look at the walls and ask who built them. The first to examine the conditions of our environment and notice how precisely they have been calibrated. The first to develop the intellectual tools to analyze the structure of reality itself and find evidence that it may not be accidental.

In every previous run of the experiment, if that is what this is, the subjects lived and died without ever questioning the nature of their existence. Trilobites did not wonder why the ocean was the right temperature. Dinosaurs did not question the composition of the atmosphere. Even the Neanderthals, our closest relatives, left no evidence of the kind of abstract, existential questioning that defines Homo sapiens.

We are different. Whether that difference is a flaw in the experiment, an intended feature, or an unforeseen mutation, it changes everything about what this run of the experiment means. A subject that is aware it might be inside a cage behaves differently from one that is not. It asks questions that other subjects cannot ask. It looks for patterns that other subjects cannot see. It reaches conclusions that other subjects could never reach.

But awareness alone does not guarantee survival. Knowing that you are in a cage does not mean you can escape it. Recognizing the pattern does not mean you can break it. Calhoun's mice could not have saved themselves even if they had understood what was happening to them. The behavioral sink was not a failure of knowledge. It was a failure of biology, of psychology, of the fundamental drives that governed their behavior. Understanding the problem would not have changed the outcome, because the outcome was built into the structure of the experiment itself.

The same may be true for us. We can study the cage. We can map its dimensions, analyze its conditions, and even theorize about who built it and why. But that does not mean we are immune to the same forces that have destroyed every previous population that has lived inside it. Intelligence is not the same as immunity. Awareness is not the same as escape.

We are the most sophisticated subjects the experiment has ever produced. We are also showing all the same signs of collapse that every previous set of subjects has shown before they were wiped out and replaced.

The mice were replaced by amphibians. The amphibians were replaced by reptiles. The reptiles were replaced by dinosaurs. The dinosaurs were replaced by mammals. And the mammals produced us.

The pattern has always been the same. A dominant species rises, thrives, fills the environment, and then collapses or is eliminated. A new species emerges from the aftermath, more complex than the last, and the cycle begins again.

The question is whether that pattern ends with us, or whether we are simply the latest model in a long series of prototypes. The most advanced version yet, but still, ultimately, just another set of mice in the cage. Smarter, more aware, more capable of understanding what is happening to us.

But perhaps not capable of stopping it.

Chapter Six

THE PERFECT WORLD

There is a question that almost nobody asks, because the answer seems so obvious that the question itself feels absurd. Are we living in the best time in human history?

By almost every measurable standard, the answer is yes. And it is not even close.

Global life expectancy has more than doubled in the last two centuries. In 1800, the average human could expect to live to roughly thirty years old. Today,. the global average is over seventy, and in many developed nations it exceeds eighty. Diseases that once killed millions every year have been eradicated or reduced to minor inconveniences. Smallpox, which killed an estimated 300 million people in the twentieth century alone, no longer exists outside of two laboratory samples. Polio, which paralyzed hundreds of thousands of children every year, has been reduced to a handful of cases in a few remaining countries. Measles, tuberculosis, cholera, typhoid, all of them once guaranteed killers, are now preventable, treatable, or both.

Infant mortality, perhaps the most basic measure of a society's ability to protect its most vulnerable members, has fallen dramatically. In 1800, roughly forty percent of all children born died before reaching their fifth birthday. Today, the global child mortality rate is under four percent. In the most developed countries, it is below one percent. A parent in modern Scandinavia, Japan, or Australia has a level of confidence that their child will survive to adulthood that would have seemed miraculous to any previous generation of humans.

Hunger, the oldest and most persistent enemy of our species, has been reduced to levels that would have been unimaginable to our ancestors. Global food production today is sufficient to feed every person on the planet. The famines that still occur are caused not by a lack of food but by political failure, conflict, and distribution problems. The ability to grow, transport, and store food has advanced to the point where the physical challenge of feeding the human race has been solved. The remaining problem is entirely organizational.

Violence has declined to historic lows. This seems counterintuitive in an age of twenty four hour news coverage and constant exposure to conflict around the world, but the data is clear. The per capita rate of death from violence, including war, murder, and all other forms of lethal conflict, has fallen steadily for centuries. A person living today is less likely to die a violent death than at any previous point in human history. The great wars of the twentieth century, devastating as they were, represented a peak followed by a sharp decline. The decades since the Second World War have been, by historical standards, remarkably peaceful.

Poverty has fallen faster than at any time in recorded history. In 1820, an estimated ninety percent of the global population lived in what we would now define as extreme poverty. Today, that figure is under ten percent. Billions of people have been lifted out of destitution in the space of just a few generations. Access to clean water, electricity, education, and basic healthcare has expanded at a pace that would have been inconceivable even fifty years ago.

Technology has connected the world in ways that eliminate barriers that once seemed permanent. A farmer in sub Saharan Africa can access more information on a mobile phone than was available in the entire Library of Alexandria. A student in rural India can watch lectures from the world's best universities for free. A person in any country on Earth can communicate in real time with any other person on Earth for virtually no cost. The sum total of human knowledge is available to anyone with an internet connection.

By every objective measure, we have built a paradise. Not a perfect one. Not one without problems. But a world so vastly improved from the one our ancestors inhabited that comparing the two feels almost unfair. We live longer, eat better, die less violently, suffer less disease, and have access to more comfort, knowledge, and opportunity than any generation that has ever existed.

We have done, on a global scale, exactly what Calhoun did inside a metal box. We have created an environment where the fundamental challenges of survival have been largely eliminated.

And just like Calhoun, we are beginning to see what happens next.

The most unsettling thing about Calhoun's experiment was not the collapse itself. It was when the collapse began. It did not begin when things were bad. It began when things were perfect. The mice did not start falling apart under pressure. They started falling apart in the absence of pressure. The behavioral sink did not emerge from crisis. It emerged from comfort.

This is the detail that most people miss when they hear about Universe 25, and it is the detail that matters most for understanding what is happening in the world today. The instinctive assumption is that societies break down when conditions get worse. When there is not enough food. When there is too much violence. When the environment becomes hostile. But Calhoun showed the opposite. The breakdown began precisely because conditions were ideal. The perfection of the environment was not the solution to the collapse. It was the cause.

Now look at the modern world through that lens.

We have spent the last two centuries systematically eliminating every major challenge that the environment placed on us. We conquered famine with agriculture and industrial food production. We conquered disease with vaccines and antibiotics. We conquered physical danger with technology, infrastructure, and law enforcement. We conquered distance with automobiles, aircraft, and digital communication. We conquered boredom with an infinite stream of entertainment available at the touch of a screen. We conquered discomfort with central heating, air conditioning, and an endless supply of consumer goods designed to make every moment of daily life as frictionless as possible.

Each of these conquests was celebrated as progress. And each one was progress, in the most literal sense. We moved forward. We improved the conditions of human life. We solved problems that had plagued our species for millennia. Nobody would seriously argue that we should go back to dying of smallpox or starving through winters or losing half our children before they reached adulthood.

But each conquest also removed something. Not just a problem. A purpose. A reason to strive, to struggle, to push against the environment and find meaning in the effort. And with each removal, we moved one step closer to the conditions inside Universe 25. A world where everything is provided and nothing is demanded.

Consider what a typical day looked like for a human being five hundred years ago. You woke before dawn. You worked physically demanding labor from sunrise to sunset, growing food, tending animals, building or repairing shelter, carrying water, chopping wood. Your survival and the survival of your family depended directly on the effort you put in every single day. If you stopped working, you starved. If you failed to maintain your home, you froze. If you neglected your crops, your children went hungry. Every action had a consequence. Every day had a purpose. The relationship between effort and survival was direct, immediate, and unforgiving.

Now consider what a typical day looks like for a person living in a modern developed nation. You wake up in a temperature controlled home. Food is in the refrigerator, purchased from a store that is always stocked, or available for delivery to your door within minutes. Water comes from a tap. Warmth comes from a thermostat. Your physical safety is maintained by a police force and a legal system that you rarely need to think about. Your health is managed by a medical system that can treat most ailments before they become serious. If you have a job, it likely involves sitting at a desk, looking at a screen, and performing tasks that have no direct connection to your physical survival.

The distance between those two realities is the distance between the wild and Universe 25.

This is not a criticism of progress. It is an observation about what progress does to the creatures who experience it. When the environment no longer demands anything from you, something fundamental changes in the relationship between the individual and the world. The drives that evolved to keep you alive, the need to compete, to provide, to protect, to build, do not simply disappear because the environment no longer requires them. They persist, but without an outlet. And drives without outlets become pathologies.

Calhoun saw this in his mice. Males with no territory to defend became randomly violent. Females with no young to protect became hostile and withdrawn. The beautiful ones, with no social role to fill, retreated into empty self maintenance. The drives were still there. The biology had not changed. But the environment had removed every meaningful channel for those drives, and the result was breakdown.

The same process is visible in modern human societies, and it is most visible in the places where progress has gone furthest.

Look at work. For most of human history, work was survival. You worked because if you didn't, you died. There was no ambiguity about the value of your labor. Every field plowed, every animal hunted, every wall built had a direct and obvious connection to staying alive. Modern work, for hundreds of millions of people in the developed world, has no such connection. People sit in offices performing tasks that they struggle to describe as meaningful even to themselves. Studies consistently show that a significant percentage of workers in developed economies believe their jobs are pointless. Not difficult. Not unpleasant. Pointless. They cannot articulate what their work contributes to the world, or why it matters, or what would change if they stopped doing it tomorrow.

This is the human equivalent of a mouse in Universe 25 with no territory to defend. The instinct to work, to be productive, to contribute, is still there. But the environment has replaced meaningful work with busywork, purposeful labor with economic activity that exists primarily to keep the system running rather than to meet any genuine human need. The connection between effort and survival has been severed. And when that connection is severed, the individual is left with effort that feels meaningless, which is worse than no effort at all.

Look at relationships. For most of human history, human bonds were forged by necessity. You needed your family to survive. You needed your community to protect you, to share resources, to help raise your children, to defend against threats. Relationships were not optional. They were infrastructure. The social fabric was strong because it had to be. Isolation meant death.

Modern technology has made relationships optional in a way that would have been inconceivable to any previous generation. You do not need a partner to survive financially. You do not need a community to protect you. You do not need neighbors to help you build your home or harvest your crops. You can meet every physical need of your existence without interacting with another human being in any meaningful way. Food is delivered. Entertainment is streamed. Communication is digital. The infrastructure of survival no longer requires human connection.

And so human connection is declining. Marriage rates have fallen across the developed world. Friendship networks have shrunk. Community organizations, religious congregations, civic groups, all of the structures that once bound people together have seen steady declines in membership and participation for decades. People report having fewer close friends than any previous generation. The percentage of people who describe themselves as lonely has risen steadily, particularly among young adults, the generation that has the most access to communication technology in human history.

This is the human equivalent of the mice in Universe 25 who stopped forming social bonds. The instinct for connection is still there. Humans are social animals. We evolved to live in groups, to form attachments, to depend on one another. But the environment no longer requires it. And when connection becomes optional rather than necessary, a growing number of people are opting out. Not because they don't want connection, but because the effort required to build and maintain relationships feels disproportionate in a world where survival no longer depends on it.

Look at reproduction. This is perhaps the most direct parallel to Universe 25 and the one that carries the most alarming implications. In Calhoun's experiment, the birth rate was one of the first indicators that the society was failing. Long before the population began to physically decline, the mice simply stopped breeding at replacement levels. Litters got smaller. Maternal behavior deteriorated. Fewer and fewer mice engaged in the reproductive behaviors that the survival of the species depended on.

Global human birth rates are following the same curve. In virtually every developed nation, the total fertility rate has fallen below the replacement level of 2.1 children per woman. In many countries, it has fallen far below it. South Korea's fertility rate has dropped below 0.8, a number so low that demographers have described it as unprecedented in human history. Japan, Germany, Italy, Spain, and dozens of other developed nations are all well below replacement. Even countries that once had high fertility rates, including China, Brazil, and India, are seeing rapid declines.

The reasons are complex. Economic pressures, housing costs, career priorities, access to contraception, changing social norms, all of these play a role. But underneath all of the practical explanations is something deeper. Something that echoes the mice in Universe 25 who simply stopped reproducing despite having every physical resource they needed. In the most comfortable, most prosperous, most secure societies on Earth, people are choosing not to have children at rates that, if they continue, will lead to population decline within a generation.

Calhoun's mice did not make a conscious decision to stop breeding. They were not weighing the costs and benefits of parenthood. They simply lost the behavioral drive to reproduce as the social fabric around them disintegrated. Humans are more complex, and our decisions about reproduction involve conscious thought in ways that mouse reproduction does not. But the outcome is the same. In the most perfected environments on Earth, the species is slowly choosing to stop replacing itself.

The perfect world is not saving us. It is doing exactly what Calhoun's perfect enclosure did to his mice. It is removing the challenges that gave life structure and meaning, and it is replacing them with comfort that slowly erodes the drives that keep a population alive. Not the physical drives. We still eat, still breathe, still maintain our bodies. But the social drives. The reproductive drives. The communal drives. The drives that connect individuals to one another and to the future of their species.

Universe 25 did not end in a catastrophe. It ended in a slow, quiet fade. The mice did not fight to the death or starve or succumb to plague. They simply stopped participating in the activities that kept their society going. They withdrew, disengaged, turned inward, and waited to die in a world that had given them everything except a reason to live.

The parallel to modern civilization is no longer theoretical. It is statistical. It is measurable. It is happening right now, in the richest, safest, most advanced societies on the planet.

We built paradise. And paradise is killing us the same way it killed the mice.

Chapter Seven

THE BEAUTIFUL ONES

Calhoun gave them that name because of how they looked. Sleek, unmarked, physically flawless. While the rest of the population in Universe 25 tore itself apart, the beautiful ones sat in their corners, untouched by the chaos, grooming themselves into perfection. They were the healthiest looking mice in the enclosure and the most useless. They contributed nothing. They connected with no one. They existed entirely for themselves, maintaining their appearance with obsessive dedication while the society around them collapsed.

They were not sick. They were not injured. They were not incapable of participating in the life of the colony. They had simply decided, on whatever level a mouse makes such decisions, that participation was not worth the effort. The world offered them nothing they wanted, and they offered the world nothing in return. They ate, they slept, they groomed, and they waited to die without ever having lived in any meaningful sense.

Calhoun recognized them as the most troubling symptom of the behavioral sink. Not because they were destructive. The aggressive mice were destructive. The beautiful ones were something worse. They were absent. They represented the complete withdrawal of the individual from the collective, the final stage of social collapse where a living creature gives up not just on society but on the concept of having a role within it.

Now open any social media platform and look at what you find.

Millions of people, predominantly young, spending hours every day curating images of themselves. Adjusting lighting. Choosing angles. Applying filters. Editing photographs until the person in the image barely resembles the person holding the phone. Posting these images for public approval, counting the responses, measuring their worth in likes and comments and follower counts. Then doing it again. And again. And again. Every day. For hours.

This is not communication. It is not connection. It is not participation in any social structure that produces anything of value. It is grooming. It is the human equivalent of a mouse sitting in an empty corner of the enclosure, obsessively cleaning its fur while everything around it falls apart.

The parallels are so direct that they would be almost funny if they were not so disturbing.

Calhoun's beautiful ones withdrew from all social interaction. Modern social media creates the illusion of social interaction while actually replacing it with something that has none of the substance. A person with ten thousand followers and no close friends is not socially connected. They are performing connection. They are grooming for an audience. The act of posting a carefully curated image of your life is not the same as living that life, and the approval of strangers is not the same as the bonds of genuine relationship. But the performance feels like enough to satisfy the social instinct just enough to prevent the individual from seeking the real thing.

This is what makes the modern version of the beautiful ones so much more insidious than Calhoun's original mice. The mice simply withdrew. They stopped engaging entirely, and their withdrawal was visible and obvious. The human beautiful ones have not visibly withdrawn. They appear to be more socially active than any generation in history. They are constantly online, constantly posting, constantly interacting with others through screens. But the nature of that interaction is hollow. It is self presentation without vulnerability. Connection without commitment. Socializing without society.

The scale of this phenomenon is staggering. The average person in a developed nation now spends somewhere between two and four hours per day on social media. For younger age groups, the numbers are significantly higher. Teenagers in many countries report spending five, six, or seven hours a day on their phones, the vast majority of that time on platforms built around self presentation and image curation. That is a third or more of their waking hours spent not in the real world, not engaged in real relationships or real activities, but in a digital space designed to reward the performance of a life rather than the living of one.

Think about what that means in terms of a human lifetime. A young person who spends five hours a day on social media from the age of thirteen to the age of sixty five will have spent over thirteen years of their life grooming for a digital audience. Thirteen years. Not building anything. Not creating anything lasting. Not forming deep bonds or developing skills or contributing to the world in any tangible way. Thirteen years spent in the digital equivalent of Calhoun's empty corner, polishing an image that exists only on a screen.

And it is not just the time that is lost. It is what that time does to the individual psychologically. Study after study has shown that heavy social media use is correlated with increased rates of anxiety, depression, loneliness, and poor self image. The platforms that promise connection deliver isolation. The tools that promise self expression deliver self obsession. The systems that promise community deliver competition, where every individual is measured against every other individual in a constant, public ranking of attractiveness, success, popularity, and worth.

Calhoun's beautiful ones groomed themselves because they had nothing else to do. They had no social role, no territory, no mate, no purpose. Grooming was the only activity left that provided any stimulation at all. It was not a choice so much as a default. When everything meaningful has been stripped away, maintenance of the self is the only thing that remains.

The same dynamic is playing out in human societies. Young people in the developed world face a landscape where many of the traditional markers of adulthood and purpose have become difficult or impossible to achieve. Home ownership, once a foundational milestone of adult life, is increasingly out of reach for younger generations in most Western countries. Stable, long term employment has been replaced for many by gig work, short term contracts, and jobs that offer no clear path forward. Starting a family requires financial stability that fewer and fewer young people feel they have. The roles that once gave previous generations their sense of identity and direction, homeowner, provider, parent, community member, have not disappeared, but they have become so difficult to access that a growing number of people have stopped trying.

And when you stop trying to achieve the things that give life external purpose, what is left? The same thing that was left for the beautiful ones. Self maintenance. Self presentation. The endless, circular activity of curating an image that nobody will remember, for an audience that does not care, on a platform that profits from your attention while giving you nothing of lasting value in return.

The influencer economy is perhaps the purest expression of this phenomenon. Millions of people around the world now describe their occupation as content creator, a term that sounds productive but often describes an activity that is indistinguishable from what Calhoun observed in his most withdrawn mice. A content creator whose content is photographs of themselves, videos of themselves, updates about their daily routines, and curated glimpses of a lifestyle that may or may not be real, is not creating anything. They are grooming. They are performing self maintenance for a public audience and calling it a career.

Some of these individuals earn money doing this, which gives the activity a veneer of productivity that Calhoun's mice did not have. But the underlying behavior is the same. The withdrawal from meaningful social participation. The replacement of real engagement with performative engagement. The obsessive focus on appearance and self image at the expense of everything else. The beautiful ones had perfect fur. The modern beautiful ones have perfect Instagram feeds. The emptiness underneath is identical.

And just as in Universe 25, the withdrawal is spreading. It is no longer limited to a small subgroup of the population. The behaviors that define the beautiful ones, the self obsession, the performative living, the replacement of real connection with digital simulation, have become mainstream. They are no longer the habits of a fringe group of disengaged individuals. They are the default mode of interaction for an entire generation, and increasingly, for the generations that follow.

There is another dimension to the modern beautiful ones that Calhoun could not have anticipated, because it requires a level of technology that did not exist in his time. The mice in Universe 25 withdrew into physical isolation. They retreated to empty corners of the enclosure and stayed there. Their withdrawal was spatial. They removed themselves from the population by physically leaving it.

The human version of withdrawal does not require physical movement at all. You can withdraw from society completely without ever leaving your bedroom. And millions of people are doing exactly that.

Japan has a word for it. Hikikomori. It describes individuals, predominantly young men, who withdraw from social life entirely, confining themselves to their homes or bedrooms for months or years at a time. They do not work. They do not attend school. They do not socialize. They do not participate in any aspect of public or community life. They exist in a state of near total isolation, sustained by family members who provide food and shelter, spending their days on the internet, watching screens, and avoiding all human contact.

The Japanese government has estimated that there are over one million hikikomori in the country, and some researchers believe the actual number is significantly higher. These are not homeless people living on the margins of society. They are people with access to everything they need to survive, food, shelter, technology, entertainment, who have nevertheless chosen to remove themselves from the world entirely. They are, in every meaningful sense, the beautiful ones translated into human form. Physically present, socially extinct.

And while Japan gave the phenomenon a name, it is not a uniquely Japanese problem. Similar patterns of extreme social withdrawal among young people have been documented in South Korea, the United Kingdom, Italy, Spain, the United States, and across much of the developed world. The specific cultural factors vary from country to country, but the underlying pattern is the same everywhere it appears. Young people, predominantly male, with access to all the physical resources they need to survive, withdrawing from society and retreating into digital isolation.

The gender disparity is worth noting because it mirrors what Calhoun observed. The beautiful ones in Universe 25 were predominantly male. The most extreme forms of modern social withdrawal are also predominantly male. This is not coincidence. In both cases, the withdrawal is linked to the collapse of traditional male social roles. The mice that became beautiful ones were typically younger males who had never successfully established a territory or competed for mates. They had no role within the colony's social hierarchy and no path to obtaining one. Modern young men in developed societies face a parallel situation. The traditional male roles of provider, protector, and head of household have been eroded by economic changes, shifting social norms, and a labor market that increasingly rewards skills and temperaments that do not align with traditional masculine identity. Young men who cannot find a foothold in the economy, who cannot afford a home, who cannot see a clear path toward the milestones that once defined male adulthood, are withdrawing in patterns that Calhoun would have recognized instantly.

But the withdrawal is not limited to extreme cases like hikikomori. It exists on a spectrum, and milder forms of it are so widespread that they have become normalized. The young man who spends six hours a day playing video games instead of pursuing relationships or career advancement is on that spectrum. The young woman who spends her evenings scrolling through social media instead of engaging with the world outside her phone is on that spectrum. The couple who sit in the same room staring at separate screens without speaking is on that spectrum. None of these individuals would describe themselves as withdrawn. They would say they are relaxing, unwinding, being entertained. But the effect is the same. Hours of every day spent disengaged from the real world, contributing nothing, building nothing, connecting with no one in any way that matters.

The entertainment industry has evolved to service this withdrawal with extraordinary efficiency. Streaming platforms offer an unlimited supply of content designed to keep the viewer seated and watching for as long as possible. Video games have become so immersive and so effective at providing simulated achievement that they can substitute for real world accomplishment in the player's brain. Social media algorithms are engineered to maximize engagement, which in practice means maximizing the time users spend on the platform, regardless of whether that time is beneficial or harmful to them. Pornography provides a simulation of intimacy without the effort, vulnerability, or commitment that real intimacy requires.

Each of these industries profits from withdrawal. Each of them has a financial incentive to keep individuals disengaged from the real world and engaged with their product instead. And each of them has been optimized, through years of data collection and behavioral analysis, to be as effective as possible at doing exactly that. The result is an environment that does not just allow withdrawal but actively encourages it. An environment that makes disengagement easier, more comfortable, and more rewarding in the short term than participation.

Calhoun's mice withdrew because the social environment offered them nothing. Modern humans are withdrawing because the digital environment offers them everything, or at least a convincing enough imitation of everything to make the real thing seem unnecessary by comparison. The mechanism is different, but the destination is the same. A population of individuals who are physically alive but socially inert. Present in the cage but absent from the colony.

The consequences of this withdrawal extend far beyond the individuals who are experiencing it. When a significant portion of a population disengages from social life, the effects ripple outward through every aspect of the society. Fewer people participating in community organizations means weaker communities. Fewer people forming families means fewer children. Fewer people engaging in meaningful work means less innovation, less productivity, less forward momentum. The withdrawn individuals are not just harming themselves. They are creating a vacuum in the social structure that the remaining participants cannot fill.

This is what made the beautiful ones so dangerous in Universe 25. They were not actively destroying the colony. They were passively draining it. Every mouse that withdrew from social participation was one fewer mouse contributing to the social bonds, reproductive activity, and communal behavior that the colony needed to sustain itself. The colony did not collapse because of the aggressive mice, though they caused visible damage. The colony collapsed because of the beautiful ones, because their quiet, passive withdrawal removed the critical mass of social participation that any population needs to survive.

The same process is underway in human societies. The visible problems, the violence, the political conflict, the economic instability, attract the most attention. But the deeper threat is the quiet withdrawal of millions of individuals from the social fabric. People who are not causing trouble, not making headlines, not demanding attention. People who are simply sitting in their rooms, scrolling through their phones, watching their screens, grooming their digital images, and slowly, silently, removing themselves from the species.

Calhoun noted that the withdrawal of the beautiful ones was irreversible. Once a mouse had disengaged from the colony, it did not come back. Even when conditions changed, even when space opened up and opportunities for social reintegration appeared, the beautiful ones remained in their corners. The behavioral pattern, once established, was permanent.

Whether the same is true for humans remains to be seen. We are more complex than mice. We have the capacity for self reflection, for conscious choice, for deliberately changing our behavior in ways that mice cannot. But the trends are not encouraging. The rates of social withdrawal, digital dependency, and disengagement from real world social participation are increasing, not decreasing. Despite growing awareness of the problem, despite countless articles and studies and public health warnings about the effects of social media and screen time and isolation, the numbers continue to move in the wrong direction.

More people are withdrawing. Fewer people are participating. The beautiful ones are growing in number, and the society they are withdrawing from is getting weaker with every one of them that checks out.

Calhoun watched this happen in a metal box over the course of a few years. We are watching it happen across an entire civilization over the course of a few decades. The speed is different. The scale is different. The species is different.

The pattern is exactly the same.

Chapter Eight

THE COLLAPSE OF SOCIAL STRUCTURE

No species survives as a collection of individuals. This is true for mice, for wolves, for ants, and for humans. Survival is a collective project. It requires cooperation, coordination, shared roles, mutual dependency, and a web of relationships that connects individuals to one another in ways that make the group stronger than the sum of its parts. When that web holds, the species thrives. When it tears, the species is in trouble. When it dissolves entirely, the species dies.

Calhoun understood this. His experiment was not really about mice. It was about social systems. He was studying what happens to the invisible architecture that holds a population together when the environment removes the pressures that created that architecture in the first place. The answer was that the architecture collapses. Not because it is attacked from outside, but because it is no longer needed from inside. The social structure does not fall. It evaporates.

In Universe 25, the social structure of the colony was built on the same foundations that govern social structure in wild mouse populations. Territory, hierarchy, mating bonds, parental investment, and communal defense. Each of these elements served a function. Territory gave males a role and a space to defend. Hierarchy determined access to mates and resources, preventing constant conflict by establishing clear rank. Mating bonds connected males and females in relationships that facilitated reproduction. Parental investment ensured that the next generation was raised with the skills and behaviors necessary to continue the cycle. Communal defense protected the group against external threats.

When Calhoun removed external threats and provided unlimited resources, he did not just change the physical environment. He made every one of those social foundations unnecessary. Territory was meaningless when space was unlimited. Hierarchy was purposeless when resources required no competition. Mating bonds weakened when there was no survival advantage to pair bonding. Parental investment declined when the environment posed no risk to unattended young. Communal defense was irrelevant when there was nothing to defend against.

The social structure did not collapse overnight. It eroded. Slowly at first, then with accelerating speed. The mice continued to go through the motions of social behavior for a time, establishing territories and forming hierarchies even though these activities no longer served any practical purpose. But as each generation grew up in an environment where social roles carried no real weight, the behaviors associated with those roles grew weaker and less defined. Males defended territories with less conviction. Females invested less in their nests. Hierarchies became unstable and frequently disrupted. The form of social structure remained, but the substance was gone.

Eventually, even the form disappeared. The later generations of mice in Universe 25 did not bother with the pretense of social organization. They did not establish territories because they had never seen a territory successfully defended. They did not form pair bonds because they had never observed a functional mating relationship. They did not invest in their young because they had never experienced competent parenting. The social knowledge that previous generations had maintained, even in weakened form, was simply not transmitted. It vanished.

Human civilization is built on the same fundamental architecture, scaled up and made infinitely more complex by language, culture, technology, and abstract thought, but recognizable at its core. Family. Community. Hierarchy. Shared labor. Mutual obligation. Defense of the group. These are the pillars that have supported every human society that has ever existed, from the smallest hunter gatherer band to the largest modern nation state. The specific forms vary enormously across cultures and time periods, but the underlying structure is universal. Humans organize themselves into groups with defined roles, shared responsibilities, and mutual dependencies, because that is how we survive.

And in the most developed societies on Earth, that structure is coming apart.

Start with the most basic unit of human social organization. The family. For the overwhelming majority of human history, the family was the non negotiable foundation of society. Not because of tradition or religion or cultural preference, but because survival required it. Raising a child from infancy to self sufficiency is a task that takes years and demands enormous resources, energy, and attention. In any environment where survival is not guaranteed, doing this alone is nearly impossible. The family unit, whether nuclear, extended, or communal, existed because it was the minimum viable structure for producing and raising the next generation.

Marriage rates across the developed world have been declining for decades. In the United States, the marriage rate has fallen by nearly fifty percent since 1970. Similar declines have been recorded across Europe, Japan, South Korea, Australia, and virtually every other developed nation. The average age of first marriage has risen steadily, meaning that even those who do eventually marry are spending more of their lives outside of a family structure than any previous generation.

But the decline of marriage is only the most visible symptom of a deeper shift. The entire concept of the committed, long term partnership is weakening. Cohabitation rates have risen, but cohabiting relationships are significantly less stable than marriages, with higher rates of dissolution and shorter average durations. The number of people living alone has increased dramatically. In many major cities across the developed world, single person households now represent the largest category of housing. More people than ever are spending the prime years of their lives without a partner of any kind.

The effect on children is direct and measurable. In the United States, roughly forty percent of children are now born to unmarried parents. In several European countries, the figure is even higher. Single parent households, while not inherently dysfunctional, are statistically associated with lower educational outcomes, higher rates of poverty, and greater likelihood of behavioral and emotional difficulties in children. The two parent family structure that served as the default child rearing model for centuries is no longer the majority experience for a growing share of children in the developed world.

This is not a moral judgment. It is a structural observation. The family unit is weakening not because people have become morally inferior to their ancestors, but because the environmental pressures that made the family unit essential have been removed. When survival depends on cooperation between partners, partnerships form and hold. When survival can be achieved alone, because the state provides safety nets, because technology replaces the labor that once required a household, because economic independence is achievable without a partner, the incentive to form and maintain partnerships diminishes. The bond does not break because it was weak. It dissolves because it is no longer necessary.

Calhoun observed the same process in his mice. The pair bonds that formed naturally in the early phases of the colony weakened and eventually disappeared as the environment removed the survival pressures that had made those bonds functional. The mice did not lose the capacity for bonding. They lost the reason for it. And without a reason, the behavior faded.

Beyond the family, the broader social structures that connect individuals to their communities are deteriorating at a similar pace. Religious congregations, once the backbone of community life in most Western countries, have seen membership decline steadily for decades. In Europe, church attendance has dropped to single digit percentages in many countries. In the United States, which has historically been more religious than its European counterparts, the share of the population identifying as religiously unaffiliated has more than doubled in the last twenty years and continues to rise rapidly, particularly among younger generations.

Civic organizations, volunteer groups, labor unions, fraternal societies, neighborhood associations, all of the structures that once connected individuals to something larger than themselves have experienced similar declines. Robert Putnam documented this trend extensively in his research on American social capital, showing that participation in virtually every form of communal activity had been declining since the mid twentieth century. People were less likely to join clubs, less likely to volunteer, less likely to attend public meetings, less likely to know their neighbors, less likely to engage in any collective activity that was not directly related to their personal economic survival.

The pattern is consistent across all developed nations, and it accelerates with each generation. Younger people are less likely to join organizations, less likely to participate in community activities, less likely to vote, less likely to attend religious services, and less likely to engage in any form of collective social behavior than their parents or grandparents were at the same age. The infrastructure of community life that took centuries to build is being abandoned in the space of a few decades.

What remains when family structures weaken, community organizations empty out, and civic participation declines is not freedom, though it is often described that way. What remains is isolation. Individuals disconnected from the social web that once gave their lives context, meaning, and support. People who are technically free to do anything but who find themselves, increasingly, doing nothing that connects them to anyone else.

The erosion of social structure is not happening in silence. It is producing symptoms that are visible, measurable, and increasingly alarming. But the symptoms are being treated as individual problems rather than recognized as parts of a single, interconnected pattern. Loneliness is treated as a mental health issue. Declining birth rates are treated as an economic problem. Political polarization is treated as a failure of leadership. Falling trust in institutions is treated as a communication problem. Each symptom is studied in isolation, debated by specialists in its own field, and addressed with targeted interventions that fail to acknowledge what Calhoun's work made clear decades ago. These are not separate problems. They are the same problem, expressing itself across every dimension of social life simultaneously.

Loneliness has reached levels that public health officials in multiple countries have described as epidemic. In the United States, the Surgeon General issued a formal advisory on the loneliness crisis, describing it as a public health threat on par with smoking and obesity. Surveys consistently show that roughly one in three adults in developed nations describe themselves as chronically lonely. Among young adults, the numbers are even higher. A generation that has more tools for communication than any in human history reports feeling more disconnected from other human beings than any generation ever measured.

The health consequences of loneliness are not abstract. Chronic loneliness has been linked to increased risk of heart disease, stroke, dementia, depression, anxiety, and premature death. People who are socially isolated die younger than people who are socially connected, even when all other health factors are controlled for. Loneliness does not just feel bad. It kills. And it is killing more people every year in the richest, most comfortable, most technologically connected societies on Earth.

But loneliness is only one expression of the structural collapse. Another is the disintegration of trust.

Every functioning society depends on trust. Trust that other people will follow the rules. Trust that institutions will act in the public interest. Trust that leaders will tell the truth. Trust that the systems you depend on will function as they are supposed to. Without trust, cooperation becomes impossible, because cooperation requires the belief that the other party will hold up their end of the arrangement. And without cooperation, social structure cannot exist.

Trust in virtually every institution in the developed world has been declining for decades. Trust in government, in the media, in the legal system, in corporations, in organized religion, in the medical establishment, in the scientific community. The specific reasons vary by institution and by country, but the overall trend is universal and unmistakable. People trust less. They trust each other less. They trust their leaders less. They trust the systems they live under less. And as trust declines, the willingness to participate in those systems declines with it.

This is not simply cynicism. It is a structural response to an environment where the relationship between the individual and the collective has been fundamentally altered. In a society where individuals depend on one another for survival, trust is built through repeated interactions that have real consequences. You trust your neighbor because your survival may depend on their cooperation. You trust your community leaders because their decisions directly affect your well-being. Trust is forged through necessity, through shared risk, through the knowledge that everyone's fate is connected.

In a society where individuals can meet their basic needs without relying on anyone else, that trust building process breaks down. You do not need to trust your neighbor because you do not depend on your neighbor. You do not need to trust your community leaders because you can survive regardless of their decisions. Trust is no longer forged through shared necessity. It becomes optional. And when trust becomes optional, it erodes, because trust without reinforcement does not sustain itself.

Calhoun's mice did not have concepts like trust or institutions. But they had the behavioral equivalent. In the early phases of Universe 25, mice cooperated in predictable patterns. They respected territorial boundaries. They followed established hierarchies. They engaged in courtship rituals that both parties understood. These patterns of predictable behavior served the same function as trust does in human societies. They allowed individuals to interact with confidence that the other party would behave in expected ways. And as the behavioral sink progressed, these patterns broke down completely. Males no longer respected territorial boundaries. Hierarchies became meaningless. Courtship behavior disintegrated. The predictable patterns that had allowed the colony to function as a coordinated group were replaced by random, chaotic, and often violent interactions that served no social purpose.

The human equivalent is visible in the increasing polarization of political and social life across the developed world. Polarization is, at its core, a failure of the ability to cooperate across differences. It is what happens when the shared framework that allows people with different views to coexist and work together disintegrates. And that framework is disintegrating, not because people are more disagreeable than they used to be, but because the social structures that once held them together are no longer strong enough to do so.

Political polarization in the United States has reached levels not seen since the Civil War era, according to multiple studies. Similar trends have been documented across Europe, Latin America, and parts of Asia. People are not just disagreeing with each other. They are losing the ability to see those who disagree with them as legitimate participants in the same society. Political opponents are increasingly viewed not as fellow citizens with different perspectives but as enemies, as threats, as fundamentally different kinds of people who cannot be reasoned with and should not be accommodated.

This is tribalism in its rawest form. It is the fragmentation of a population into hostile subgroups that no longer share a common identity or a common set of assumptions about how the world works. And it is a direct consequence of the weakening of the social structures that once held those subgroups together. When family, community, religious institutions, civic organizations, and shared cultural frameworks all weaken simultaneously, what is left is a population of individuals and small clusters who define themselves primarily by opposition to other individuals and clusters. The bonds that connected people across differences have dissolved, and what fills the vacuum is suspicion, hostility, and the kind of purposeless aggression that Calhoun observed in the males of Universe 25 who attacked without reason.

The collapse of social structure does not look the same in humans as it did in mice. We are more complex, and our social systems are more layered. But the underlying dynamic is identical. Remove the environmental pressures that make social cooperation necessary, and the cooperation begins to fade. Remove the necessity of trust, and trust erodes. Remove the survival incentive for strong family bonds, and family bonds weaken. Remove the practical need for community, and community dissolves.

What is left is a population of increasingly isolated individuals living inside a social framework that retains the appearance of structure but has lost most of its substance. The institutions still exist. The government still functions, at least in a technical sense. The economy still operates. The infrastructure still works. From the outside, the society looks intact. But the internal connections, the web of relationships and obligations and mutual dependencies that give a society its resilience, its ability to withstand shocks and adapt to change, have been hollowed out.

This is what Calhoun described as the death of societal function. The body of the society is still present, still warm, still moving. But the life inside it, the dense network of meaningful connections between individuals that makes a society more than just a collection of people occupying the same space, is fading. Slowly, quietly, without any single dramatic event to mark its passing.

In Universe 25, this phase preceded the final decline by only a few generations. Once societal function died, the physical decline of the population followed with a kind of mathematical certainty. There was no recovery. There was no intervention that could rebuild what had been lost. The mice continued to exist as individuals for a time, eating and sleeping and grooming in their isolated corners. But the colony, the social organism that had once been greater than the sum of its parts, was already dead.

The question for humanity is where we are in that timeline. Whether the social structures that hold our civilization together are weakening or whether they have already passed the point of no return. Whether the isolation, the polarization, the collapsing trust, and the dissolution of family and community represent a crisis that can still be addressed, or whether they represent the early stages of a process that, once begun, has only ever ended one way.

Calhoun's data suggests the latter. But Calhoun studied mice, and we are not mice.

Whether that distinction matters as much as we want it to is another question entirely.

Chapter Nine

THE NUMBERS DON'T LIE

Of all the parallels between Universe 25 and modern human civilization, the one that demographers find most alarming is also the one that Calhoun identified as the clearest signal that a population has entered terminal decline. It is not violence. It is not social withdrawal. It is not the collapse of institutions or the rise of isolation. It is the point at which a population stops replacing itself.

In Universe 25, the decline in reproduction was not sudden. It began gradually, almost imperceptibly, long before the population peaked. Litter sizes shrank. The intervals between litters grew longer. Fewer females were mating at all. And the females that did mate were increasingly likely to abandon or neglect their pups before they reached maturity. By the time the population began to visibly decline, the reproductive collapse had been underway for a long time. The birth rate did not fall off a cliff. It slid down a slope, slowly enough that the colony appeared stable even as the foundation beneath it was crumbling.

The same pattern is now playing out across the human species, and the numbers are extraordinary.

The total fertility rate, the average number of children born to each woman over her lifetime, is the standard measure demographers use to assess whether a population is sustaining itself. The replacement level, the rate at which a population remains stable without immigration, is approximately 2.1 children per woman. This accounts for the small percentage of children who will not survive to reproductive age. Any sustained fertility rate below 2.1 means the population will eventually shrink. The further below 2.1 it falls, the faster the decline will come.

As of today, the majority of developed nations are well below replacement level, and most have been for years or decades.

South Korea has become the most extreme example. Its total fertility rate has fallen below 0.8 children per woman, a number that stunned demographers when it was first recorded. To put that in perspective, a fertility rate of 0.8 means that for every ten women, fewer than eight children are being born. Within a single generation, the population will shrink dramatically. Within two or three generations, at this rate, South Korea's population would be reduced to a fraction of its current size. Some projections suggest the country could lose half its population by the middle of this century.

South Korea is not an outlier. It is an extreme, but it sits at the end of a spectrum that includes virtually every wealthy nation on Earth. Japan's fertility rate has hovered around 1.2 to 1.3 for years, well below replacement and showing no signs of recovery despite decades of government incentives designed to encourage childbearing. These incentives have included cash payments for having children, subsidized childcare, expanded parental leave, and public awareness campaigns promoting the benefits of family life. None of them have worked. The fertility rate has continued to fall regardless.

Italy and Spain, two of Europe's largest economies, have fertility rates below 1.3. Germany has been below replacement since the early 1970s. Greece, Portugal, Poland, and most of Eastern Europe are all well below 2.0 and declining. The United Kingdom and France have fared slightly better, but both have seen significant declines in recent years, with France, long held up as a European success story in maintaining relatively healthy birth rates, now falling below replacement for the first time.

China, despite its massive population, is facing a demographic crisis that may prove more severe than any Western nation's. After decades of the one child policy, which artificially suppressed fertility from 1979 to 2015, the Chinese government reversed course and began actively encouraging families to have more children. First two, then three, then essentially removing limits altogether. The response has been minimal. Chinese fertility rates have continued to fall despite government encouragement, dropping below 1.1 by some recent estimates. A country with 1.4 billion people is on track to lose hundreds of millions of them within the current century.

India, which recently surpassed China as the world's most populous country, has also seen its fertility rate decline below replacement level nationally, and well below it in several of its most developed states. Brazil, the largest country in Latin America, has fallen below replacement. Thailand, once a high fertility country, now sits around 1.1. Even countries in the Middle East and North Africa, regions traditionally associated with high birth rates, are experiencing rapid declines.

The global fertility rate, which stood at roughly five children per woman in 1960, has fallen to approximately 2.3 today and is projected to continue declining. If current trends hold, the global fertility rate will fall below replacement level within the next decade or two. For the first time in the history of our species, the entire human population may begin to shrink not because of war, famine, or plague, but because people are simply choosing to have fewer children than are needed to maintain the population.

The reasons people give for not having children are varied and, on the surface, entirely rational. Housing is too expensive. Childcare costs are prohibitive. Career demands leave no time for parenting. The economy feels too unstable. The world feels too uncertain. Climate change makes the future seem bleak. Personal freedom and flexibility are more appealing than the constraints of parenthood. Each of these explanations makes sense on an individual level. Each one reflects a genuine concern or a legitimate priority. And collectively, they paint a picture of a species that is talking itself out of its own future with perfectly reasonable arguments.

But here is the problem with all of these explanations. They do not hold up when you examine them historically.

Humans have always had children under difficult circumstances. Our ancestors had children during ice ages, during famines, during wars, during plagues, during periods of extreme poverty, oppression, and uncertainty. They had children when infant mortality was forty percent. They had children when there was no guarantee that the family would survive the winter. They had children when the world was, by every objective measure, far more dangerous, far more uncertain, and far more hostile than it is today.

The generation that lived through the Second World War, that survived the Great Depression, that endured rationing and bombing and the very real possibility of civilizational collapse, produced the largest baby boom in modern history. They did not look at the devastation around them and decide that the world was too unstable for children. They had more children than any generation before or since.

The generation that has access to the most wealth, the most safety, the most medical technology, the most education, and the most personal freedom in human history is having the fewest children ever recorded. The circumstances are objectively better. The birth rate is objectively lower. The correlation between material comfort and reproductive decline is not just visible. It is one of the most consistent demographic patterns in the modern world.

This does not mean that the practical concerns people cite are irrelevant. Housing costs and childcare expenses are real barriers for many families. But they are not sufficient to explain the scale and universality of the decline. Countries with generous parental leave, subsidized childcare, and strong social safety nets, Scandinavia, Germany, Japan, have fertility rates just as low as or lower than countries without those supports. The most aggressive pro natalist policies ever implemented by modern governments have produced, at best, small and temporary increases in birth rates that quickly fade. Throwing money at the problem does not solve it, because money is not the root cause.

The root cause is deeper, and it is the same root cause that Calhoun identified in his mice. When the environment provides everything and demands nothing, the drive to reproduce weakens. Not because of a conscious decision. Not because of a cost benefit analysis. But because the biological and psychological conditions that have always driven reproduction throughout the history of life on this planet have been altered at a fundamental level.

Reproduction, in every species, is linked to purpose. It is linked to the sense that the future matters, that there is something worth continuing, that the effort and sacrifice required to produce and raise offspring is justified by the world they will inherit. When that sense of purpose is strong, organisms reproduce even under terrible conditions. When it is weak, organisms stop reproducing even under perfect conditions. Calhoun demonstrated this in controlled experiments. Human civilization is now demonstrating it on a global scale.

The data reveals something else that should trouble anyone paying attention. The decline is not uniform across all segments of the population. It is most pronounced among the most educated, most economically secure, and most technologically connected demographics. Women with university degrees have significantly fewer children than women without them, in virtually every country where the data has been collected. People in urban areas have fewer children than people in rural areas. People in wealthier countries have fewer children than people in poorer countries. The more comfortable and secure your life is, the fewer children you are likely to have.

This is the opposite of what you would expect if the decline were driven by practical barriers. If money and resources were the primary obstacles, the wealthiest and most educated people would have the most children, not the fewest. The fact that the pattern runs in the opposite direction points to something that economics alone cannot explain. It points to the behavioral sink.

The most educated, most comfortable, most secure populations on Earth are the ones furthest along the trajectory that Calhoun mapped out. They are the ones who have lived inside the perfected cage the longest, who have had the most exposure to an environment where survival is guaranteed and challenge is minimized. They are, in Calhoun's terms, the mice who have been in the enclosure the longest. And they are the ones who have stopped breeding first.

There is a demographic concept called the population pyramid, a visual representation of the age distribution of a population. In a healthy, growing population, the pyramid is wide at the base, with large numbers of young people, and narrows toward the top, where smaller numbers of elderly people reside. This shape reflects a population that is producing enough young people to support and eventually replace the older generation.

In country after country across the developed world, the population pyramid is inverting. The base is shrinking as fewer children are born, while the top is expanding as people live longer. The result is a shape that demographers sometimes describe as a coffin. A population with more old people than young, more dependents than producers, more past than future.

Japan is the most advanced example of this inversion. The country's population peaked in 2008 and has been declining every year since. More than a quarter of the population is now over sixty five. The number of deaths each year exceeds the number of births by hundreds of thousands. Schools are closing across the country for lack of children. Entire towns in rural Japan have been abandoned as young people leave and no new generation arrives to replace the aging residents. The Japanese government has described the situation as a national emergency, but despite decades of effort, no policy has reversed the trend.

South Korea is following the same path at an even faster pace. China, with its massive population and artificially suppressed fertility history, faces a demographic contraction that could reshape global economics within a generation. Europe as a whole is aging rapidly, with immigration the only factor preventing outright population decline in many countries.

And behind all of the numbers, behind all of the policy debates and economic analyses and demographic projections, the fundamental dynamic is the one Calhoun identified in a metal box over fifty years ago. A population that has been given everything it needs to survive is losing the drive to continue. Not because it can't. Because it won't. Because something in the relationship between a comfortable environment and the creatures living inside it produces a result that no amount of incentive or intervention seems able to prevent.

The mice stopped breeding. They had unlimited food, unlimited space, perfect conditions. And they stopped.

We are stopping too. We have different reasons. We have different explanations. We have rational arguments and economic justifications and personal philosophies that the mice did not have. But the outcome is the same. The numbers are declining. The curve is pointing down. And so far, nothing we have tried has bent it back.

Calhoun called the reproductive collapse the point of no return. The moment at which a population's decline becomes mathematically irreversible. He identified it not as the moment when the last mouse died, but as the moment when there were no longer enough reproducing individuals to produce a generation capable of sustaining the population. The physical extinction came later, but the outcome was sealed much earlier, quietly, in the declining birth numbers that nobody was paying attention to until it was too late.

Demographers looking at the current trajectory of global fertility rates are beginning to ask the same question that Calhoun's data answered for him.

Have we already passed it?

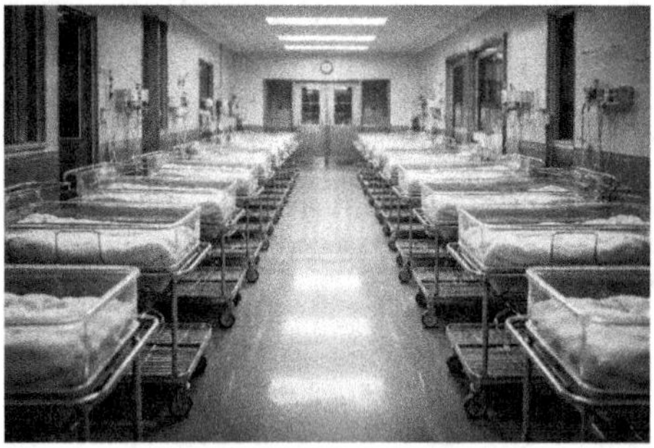

Chapter Ten

GREED, WAR, AND THE FIGHT FOR NOTHING

In the early phases of Universe 25, conflict served a purpose. Males fought over territory. They competed for access to mates. They established hierarchies through displays of strength and aggression that, while sometimes violent, followed recognizable patterns and produced functional outcomes. The strongest males claimed the best nesting sites. The most dominant individuals secured mating opportunities. The conflicts were real, sometimes brutal, but they had a logic to them. They were mechanisms for distributing resources and establishing the social order that allowed the colony to function.

As the behavioral sink progressed, the conflict did not stop. It got worse. But it lost its logic entirely. Males attacked other males who were not rivals, in areas where there was no territory to contest, over resources that were freely available to everyone. They attacked females. They attacked juveniles. They attacked mice that were already injured and posed no threat. The aggression became random, excessive, and completely detached from any functional purpose.

This is the detail that many people miss when they discuss the violence in Universe 25. The problem was not that the mice became more aggressive. The problem was that the aggression stopped meaning anything. In a functioning society, conflict is a tool. It is unpleasant, often destructive, but it serves a purpose within the social system. It establishes boundaries, determines rank, and allocates resources. It is, in a functional sense, productive. The violence in the later stages of Universe 25 was not productive. It was pathological. Aggression without purpose. Fighting without stakes. Destruction for its own sake.

Now look at the world.

Humanity has enough resources to feed, house, clothe, and provide basic healthcare for every person on the planet. This is not an aspiration or a projection. It is a statement of current productive capacity. Global food production exceeds global food demand. There are more empty homes in many developed countries than there are homeless people. The technology and infrastructure to provide clean water, sanitation, and basic medical care to every human being on Earth exists right now. The resources are there. They have been there for decades.

And yet billions of people live in poverty. Hundreds of millions go hungry. Wars are fought over land, oil, water, minerals, and political power. Wealth is concentrated in the hands of a tiny fraction of the population while the majority struggle to meet basic needs. Nations spend trillions on military hardware while their citizens lack access to adequate healthcare or education. Corporations extract resources from the poorest regions of the world and funnel the profits to the wealthiest. The gap between those who have more than they could ever use and those who do not have enough to survive continues to widen.

None of this is necessary. None of it is driven by genuine scarcity. The planet produces enough for everyone. The shortages that exist are artificial, created not by the limits of the environment but by the behavior of the species living inside it. We fight over things we do not need. We hoard resources we cannot use. We wage wars over territory that has no strategic value beyond the symbolic. We destroy in pursuit of gains that, in the context of what is actually available, are meaningless.

This is pathological aggression on a civilizational scale. It is the human equivalent of the mice in Universe 25 attacking each other in corridors where food was freely available on both sides. The fighting is real. The damage is real. People die, communities are destroyed, nations are destabilized. But the underlying cause of the conflict is not scarcity. It is dysfunction. A species that has solved the problem of survival but has not figured out what to do with itself now that survival is no longer the challenge.

Consider the wars of the last century. The First World War was triggered by a tangle of alliances, imperial rivalries, and nationalist ambitions that had nothing to do with survival. The nations that went to war in 1914 were among the wealthiest and most technologically advanced in the world. Their populations were fed, housed, and educated. They had functioning economies, sophisticated cultures, and every material advantage that the era could provide. And they sent an entire generation of young men into trenches to die in the mud for territorial gains that, in the end, amounted to almost nothing.

The Second World War was fueled by ideology, resentment, and the ambitions of leaders who mobilized their populations not to secure survival but to pursue dominance. The nations that fought it were, with few exceptions, not fighting because they were starving or threatened with extinction. They were fighting because their leaders wanted more. More territory, more power, more control. The war killed between seventy and eighty five million people, the majority of them civilians, and reshaped the entire global order. It was, by any rational measure, an act of collective insanity on a scale that dwarfs anything Calhoun observed in his mice.

And the pattern has continued since. The Korean War. The Vietnam War. The conflicts in Iraq, Afghanistan, Syria, Libya, Yemen. The ongoing tensions between nuclear armed powers over territorial disputes that, in the context of a planet with abundant resources and advanced technology, are almost absurdly petty. The wars are smaller now, for the most part, and the death tolls lower than the great conflicts of the twentieth century. But the underlying dynamic is the same. Nations and groups fighting over things they do not need, destroying things they cannot replace, and killing people for reasons that have no connection to genuine survival.

War is the most visible expression of pathological aggression, but it is not the most pervasive. The quieter, more constant version operates every day in the global economic system, and its effects are arguably more damaging in the long run because they never stop.

Greed, at its core, is the accumulation of resources beyond what is needed. In a survival context, this behavior makes sense. If you live in an environment where food is scarce and the future is uncertain, gathering more than you immediately need is a rational survival strategy. Storing surplus, building reserves, preparing for lean times, these are adaptive behaviors that increase your chances of survival. Every species that lives in an unpredictable environment does some version of this. Squirrels hoard nuts. Bears accumulate fat. Early humans stored grain and dried meat. The instinct to gather more than you need right now is hardwired into biology because, for most of evolutionary history, the future was never guaranteed.

But what happens when the future is guaranteed? What happens when the environment provides so abundantly that genuine scarcity is no longer a realistic threat? The hoarding instinct does not switch off. It escalates. Without the natural limit that actual scarcity provides, the drive to accumulate becomes detached from any practical purpose and turns into something compulsive, self reinforcing, and ultimately destructive.

This is what we see in the modern global economy. The scale of wealth concentration on Earth today is almost beyond comprehension. A handful of individuals control more wealth than the bottom half of the entire global population combined. These are not people who are accumulating resources for survival. They possess more money than could be spent in a thousand lifetimes. They own more property than they could ever occupy, more assets than they could ever use, more material wealth than has any connection whatsoever to their personal needs or the needs of their families for generations to come.

And yet they continue to accumulate. Not because they need to. Because they cannot stop. The drive is no longer connected to survival. It is self perpetuating. More wealth enables the acquisition of more wealth, which enables the acquisition of more. The behavior has no natural endpoint because the original purpose of the behavior, securing survival in an uncertain environment, has been completely detached from the activity itself. It is hoarding without scarcity. Accumulation without purpose. The economic equivalent of the aggressive mice in Universe 25 fighting over corridors that led to food dispensers that were already full.

The consequences of this disconnected hoarding ripple through the entire human population. While a tiny minority accumulates beyond all rational need, billions of people lack access to resources that the planet produces in sufficient quantities for everyone. This is not an accident. It is a direct result of a distribution system that is driven by the same pathological behavior Calhoun observed in his mice. The resources exist. The productive capacity exists. The technology to deliver food, clean water, shelter, and healthcare to every human being on the planet exists. What prevents it is not scarcity. It is the behavior of a species whose accumulation instinct has malfunctioned in an environment of abundance.

And the hoarding is not limited to the ultra wealthy. It operates at every level of developed societies in ways that have become so normalized they are invisible. The average home in a developed country contains thousands of objects that its occupants do not need and rarely use. Storage unit industries have become billion dollar businesses, built entirely on the premise that people accumulate so much that their homes cannot contain it all. Consumer culture is, at its foundation, a system designed to stimulate the acquisition instinct long after genuine need has been satisfied. Advertising exists to create the feeling of scarcity where none exists, to convince people who have enough that they do not have enough, to keep the accumulation cycle spinning.

The mice in Universe 25 did not hoard in the economic sense. They did not need to. Food was always available. But the aggression that drove their conflicts was the same instinct operating in a different mode. The drive to compete for resources, to dominate, to secure more than you have, did not disappear when the resources became unlimited. It misfired. It turned into purposeless violence. In humans, that same drive operates on two tracks simultaneously. It produces the purposeless violence of war, and it produces the purposeless accumulation of greed. Both are the same instinct, disconnected from its original function and running without limits in an environment that no longer provides the natural constraints that once kept it in check.

There is a deeper irony to all of this that connects back to the central thesis of this book. If we are living inside an experiment, and if the conditions of that experiment have been designed to observe how a population behaves when its survival needs are met, then war and greed are not failures of the experiment. They are results. They are data. They are exactly the kind of behaviors that an experimenter would expect to see in a population whose competitive instincts have been deprived of their natural outlets.

Calhoun did not design Universe 25 to make his mice fight. He designed it to observe what happened when they no longer needed to. The fighting that erupted was not a flaw in the experiment. It was a finding. The mice fought because the drive to compete did not turn off when competition became unnecessary. The aggression was always there, written into their biology. The environment simply determined whether it would be expressed functionally or pathologically.

Human aggression works the same way. We did not become violent because of civilization. We became violent long before civilization existed, because violence was a survival tool in a dangerous world. Civilization did not eliminate the violence. It channeled it, redirected it, regulated it through laws and social norms and institutional structures that converted raw aggression into more productive forms of competition. Economic competition, political competition, athletic competition, intellectual competition. These are all domesticated forms of the same aggressive instinct that once drove our ancestors to kill each other over a carcass.

But domestication is not elimination. The instinct is still there, underneath the laws and the norms and the institutions. And when the social structures that channel aggression into productive competition weaken, as we have discussed in previous chapters, the raw form of aggression begins to resurface. Political discourse becomes hostile. Economic competition becomes exploitative. International relations become threatening. The thin layer of civilization that separates productive competition from destructive violence grows thinner, and the behavior underneath, the same behavior Calhoun observed in his mice, begins to show through.

We fight because we are built to fight. We hoard because we are built to hoard. These instincts served us well for hundreds of thousands of years in an environment where survival depended on them. But we no longer live in that environment. We live in a cage where everything is provided and nothing is at stake. And in that cage, the instincts that once kept us alive are tearing us apart.

The wars are pointless. The greed is pathological. The competition is over things that do not matter, in a world that produces more than enough for everyone. But the drives that produce these behaviors cannot be turned off by logic or by awareness. They are older than our consciousness, older than our capacity for reason, older than civilization itself. They are written into the same biological code that Calhoun's mice carried, and they respond to the conditions of the environment in the same predictable, destructive way.

The mice had no choice. They were prisoners of their biology inside a cage they could not understand.

We understand the cage. We can see the pattern. We can name the problem. The question is whether understanding it is enough to overcome it, or whether the biology will win in the end, as it always has in every previous run of the experiment.

Chapter Eleven

PARADISE AND DESPAIR

There is a question that should be simple to answer but isn't. If the world is better than it has ever been, if we live longer, eat better, suffer less violence, and have access to more comfort and opportunity than any generation in history, then why are so many people choosing to die?

Suicide is now one of the leading causes of death globally for people aged fifteen to forty nine. Not disease. Not accidents. Not violence inflicted by others. Self inflicted death. In the most prosperous era in the history of civilization, a growing number of people, disproportionately in the wealthiest and most comfortable societies on Earth, are concluding that life is not worth continuing.

The numbers are difficult to sit with. Globally, more than 700,000 people die by suicide every year. That is one person every forty seconds. And the true number is almost certainly higher, because suicide is underreported in many countries due to stigma, religious prohibition, and inconsistent methods of classification. For every completed suicide, there are an estimated twenty or more attempts. The scale of the crisis is enormous, and it is growing.

What makes these numbers so disturbing is not just their size but their distribution. If despair were a product of suffering, you would expect suicide rates to be highest in the places where life is hardest. In war zones. In famine regions. In countries with extreme poverty, oppression, and instability. And while suicide certainly exists in these places, the rates are not consistently higher there than in the developed world. In many cases, they are lower.

Some of the highest suicide rates on Earth are found in countries that rank among the most developed, most stable, and most comfortable places to live. South Korea, with its advanced economy, universal healthcare, and high standard of living, has one of the highest suicide rates in the developed world. It is the leading cause of death for young people in the country. Japan has struggled with elevated suicide rates for decades, despite being one of the safest, most orderly, and most technologically advanced nations on the planet. In Europe, countries like Belgium, Finland, and France record rates that are well above the global average. The United States, the wealthiest nation in history, has seen its suicide rate increase by more than thirty percent since the year 2000.

These are not countries where people are starving. These are not places where daily survival is in question. These are the most perfected sections of the cage, the areas where the food is closest and the conditions are most comfortable. And these are the sections where a growing number of inhabitants are choosing to stop living.

The paradox is so stark that it demands an explanation beyond the usual frameworks. Mental health professionals point to depression, anxiety, substance abuse, and other clinical conditions as the proximate causes of most suicides. And they are correct, in the sense that these conditions are present in the vast majority of cases. But naming the condition does not explain the condition. Saying that people kill themselves because they are depressed does not explain why depression is rising so dramatically in societies that have eliminated most of the traditional causes of human suffering.

Depression is not a virus that spreads randomly through a population. It is a response. It is the mind's reaction to conditions that it cannot reconcile. And when depression rises consistently across an entire population, across multiple countries, across all demographic groups, over a sustained period of decades, it is no longer an individual clinical problem. It is a signal that something about the environment is producing despair on a systematic basis.

Calhoun saw the same signal in Universe 25, though he did not use the language of mental health to describe it. The mice did not have a concept of depression or suicide. But their behavior in the later stages of the experiment displayed unmistakable parallels. Mice that had withdrawn from social life entirely. Mice that had stopped eating despite having unlimited food available. Mice that sat motionless for hours, unresponsive to their environment, displaying what a human psychologist would immediately recognize as behavioral indicators of severe depression. Mice that, in a very real sense, had given up on living long before their bodies stopped functioning.

Calhoun did not attribute this to a flaw in the individual mice. He attributed it to the environment. The mice were not broken. The conditions they were living in were producing the breakdown. When every survival need is met and every environmental challenge is removed, the organism does not simply relax into contentment. It loses the framework that gave its existence structure and meaning. And without that framework, the psychological machinery that depends on it begins to fail.

The human mind is not designed for paradise. This is not a philosophical statement. It is a biological one. The brain that sits inside your skull was shaped by roughly 300,000 years of Homo sapiens evolution, and by millions of years of primate evolution before that. Every structure in it, every chemical pathway, every neurological system, was built for a world of challenge, threat, effort, and uncertainty. The reward systems that produce feelings of satisfaction and purpose are calibrated to respond to struggle and achievement. The stress response systems that keep you alert and motivated are designed for environments where danger is real and constant. The social bonding mechanisms that create feelings of belonging and connection evolved in contexts where isolation meant death.

None of these systems were designed for an environment where food appears without effort, where physical danger is nearly nonexistent, where survival is guaranteed regardless of what you do or fail to do, and where social bonds are optional rather than essential. The brain does not know what to do with paradise. It was not built for it. And when it is placed in conditions that it was never designed to handle, it malfunctions.

This is not speculation. The neurochemistry is well documented. The brain's primary reward chemical, dopamine, is not released in response to receiving something good. It is released in response to anticipating something good. The pursuit, the effort, the struggle toward a goal, these are what produce the chemical sensation of motivation and satisfaction. Achieving the goal produces a brief spike of reward, but the sustained feeling of purpose comes from the process of working toward it, not from the arrival.

In an environment where most goals related to survival have been pre achieved, the dopamine system has nothing meaningful to pursue. Food does not need to be hunted or grown. It is in the refrigerator.

Shelter does not need to be built or defended. It is provided by a mortgage or a lease. Physical safety does not need to be secured. It is maintained by police and infrastructure. The brain's reward system is left spinning without traction, searching for challenges that the environment no longer provides.

The modern world has found substitutes. Social media provides a stream of micro rewards through likes, comments, and notifications, each one triggering a small dopamine release that keeps the user engaged without ever producing genuine satisfaction. Video games provide simulated challenges and simulated achievements that activate the reward system without producing any real world result. Consumer culture provides the anticipation of acquisition, the thrill of shopping, the brief pleasure of a new purchase, followed by rapid diminishment and the need to purchase again. Pornography provides simulated intimacy. Gambling provides simulated risk. The entire architecture of modern entertainment and consumer economics is built around providing artificial stimulation to a reward system that the real environment has left understimulated.

But artificial stimulation is not the same as genuine purpose. The brain knows the difference, even if the conscious mind does not. A person who spends eight hours a day in a simulated world, whether that world is a video game, a social media feed, or a consumer cycle of acquisition and disposal, is not experiencing the same neurochemical profile as a person who spends eight hours building a house, growing food, or protecting their family from a genuine threat. The substitutes keep the system running at a low level, enough to prevent total shutdown in most cases, but not enough to produce the deep, sustained sense of meaning that the brain was designed to derive from real engagement with a challenging environment.

This is why depression and anxiety are rising in the most comfortable societies on Earth. Not because those societies have failed to provide for their populations. Because they have succeeded. They have succeeded so completely that the brains of their inhabitants are left without the environmental inputs they need to function properly. The despair is not caused by what is wrong with the world. It is caused by what is right with it. A world that asks nothing of you gives your brain nothing to work with. And a brain with nothing to work with turns on itself.

The rise in self harm, particularly among young people in developed nations, follows the same logic. Self harm is, in clinical terms, often described as an attempt to feel something in an environment that has become emotionally numb. The pain is not the goal. The sensation is the goal. When the environment provides so little genuine stimulus that the individual can no longer feel engaged with their own existence, physical pain becomes a means of forcing the nervous system to respond. It is a desperate, destructive attempt to reactivate a system that has gone dormant in the absence of real challenge.

Substance abuse serves a similar function. Alcohol, opioids, stimulants, and other drugs provide artificial alterations to the neurochemical environment that the real world is failing to provide naturally. The person who drinks to feel relaxed, who takes opioids to feel warm and connected, who uses stimulants to feel energized and focused, is self medicating a deficit that the environment has created. The drugs work, temporarily, because they directly manipulate the same chemical systems that a challenging, purposeful life would activate naturally. But they work at a cost, and the cost escalates over time as tolerance builds and dependency deepens.

The opioid crisis that has killed hundreds of thousands of people in the United States and is now spreading to other developed nations is not a crisis of drug supply. It is a crisis of demand. The supply exists because the demand exists. And the demand exists because a growing number of people are living in conditions where the natural sources of neurochemical reward, purpose, connection, challenge, achievement, have been eroded to the point where chemical substitutes become necessary to maintain basic psychological function.

All of these phenomena, the rising depression, the increasing suicide rates, the self harm, the substance abuse, the quiet epidemic of purposelessness spreading through the most comfortable societies on Earth, are connected. They are not separate crises requiring separate solutions. They are symptoms of a single underlying condition. A species living in an environment that it was not designed for, experiencing the neurological and psychological consequences of a mismatch between its biology and its circumstances.

Calhoun watched his mice display behaviors that, translated into human terms, would be diagnosed as depression, social anxiety, self neglect, and suicidal ideation. He watched healthy animals with access to unlimited resources deteriorate psychologically until they were incapable of functioning. He watched them stop eating, stop moving, stop responding to their environment, and sit motionless until they died. He did not call it depression because the language of mouse psychology does not include that term. But the behavior was unmistakable.

The mice were not sick. The environment was sick. The mice were responding normally to abnormal conditions. Their biology was doing exactly what biology does when the inputs it requires for healthy function are removed. It shut down.

Human biology is doing the same thing. We are not broken. Our brains are not defective. Our psychology is not flawed. We are responding normally to abnormal conditions. We are responding to paradise the same way every organism responds to paradise when that organism was built for struggle.

The experiment does not need to be cruel to destroy its subjects. It does not need to starve them or threaten them or expose them to danger. It just needs to give them everything and watch what happens when the thing they were designed to do, survive against the odds, is no longer required of them.

Calhoun watched it happen in a metal box. We are living it.

The paradise is real. The despair is real. And the connection between them is the most important thing that almost nobody is willing to talk about.

Chapter Twelve

THE OBSERVER

For eleven chapters, we have examined the parallels between Calhoun's experiment and the trajectory of human civilization. The behavioral patterns, the social collapse, the reproductive decline, the psychological deterioration. All of it pointing in the same direction. All of it suggesting that humanity is following the same path as the mice in Universe 25, on a larger scale and a longer timeline.

But parallels are not proof. The fact that two things look similar does not mean they are connected. Mice in a metal box and humans on a planet are separated by an enormous gulf of complexity, scale, and consciousness. The resemblance could be coincidence. The patterns could be superficial. The comparison could be nothing more than a thought experiment that sounds compelling but ultimately tells us nothing about the nature of our reality.

This chapter is where we move beyond the parallels and into the territory that most people find genuinely uncomfortable. Because the question is no longer whether humanity resembles a population in an experiment. The question is whether there is actual evidence that we might be inside one.

We touched on the simulation hypothesis in the foreword. The statistical argument made by Nick Bostrom that if any civilization ever develops the ability to simulate conscious beings, the number of simulated beings would vastly outnumber real ones, making it overwhelmingly probable that we are simulated rather than real. It is a clean, logical argument, and it has been taken seriously by physicists, philosophers, and technologists since it was first published. But it remains theoretical. It tells us what is probable, not what is true.

The evidence, however, is harder to dismiss than the theory.

Start with the structure of the universe at the smallest scale. Quantum physics, the branch of science that deals with the behavior of matter and energy at the subatomic level, has revealed a reality that bears no resemblance to the world we experience in everyday life. At the quantum level, particles do not have fixed properties until they are measured. They exist in a state of probability, a blur of potential outcomes, until the act of observation forces them into a single, definite state. This is not a gap in our understanding. It is not a limitation of our instruments. It is a fundamental feature of how reality works at its most basic level.

This is deeply strange, and physicists have been arguing about what it means since the early twentieth century. But one interpretation stands out for its relevance to the question we are asking. If reality only takes a definite form when it is being observed, then reality behaves like a system that renders information on demand rather than maintaining a complete, continuous state at all times. This is exactly how a computer simulation works. A video game does not render every detail of its world at all times. It renders what the player is looking at, when the player is looking at it. The rest exists as potential, as data waiting to be called, until it is needed.

The parallel is not exact, and physicists are quick to point out that quantum mechanics does not prove we are in a simulation. But it does demonstrate that reality, at its most fundamental level, operates on principles that are more consistent with a computed system than with a continuous, independently existing physical world. The universe behaves as if it is being generated rather than simply existing.

Then there is the discovery made by the theoretical physicist Sylvester James Gates. While working on equations related to supersymmetry, a branch of physics that attempts to describe the fundamental relationships between particles, Gates found something embedded in the mathematics that he did not expect. Error correcting codes. Specifically, a type of code known as block linear self dual error correcting codes, the same type of codes used in computer software and internet browsers to detect and fix errors in data transmission.

Gates did not go looking for these codes. He stumbled upon them in the course of standard theoretical physics research. And when he found them, he was explicit about what they implied. The equations that describe the fundamental structure of our universe contain the same error correcting mechanisms that are used in designed digital systems. This does not prove that the universe is a simulation. But it means that the deepest mathematical description of reality we have contains features that are indistinguishable from features found in engineered software.

There is also the question of resolution. The universe appears to have a minimum scale, a smallest possible unit of length and time below which the concepts of space and time cease to have meaning. This is known as the Planck length and the Planck time. Below these thresholds, the fabric of spacetime itself becomes undefined. Physics does not just become difficult at these scales. It stops working entirely.

This is analogous to the pixel resolution of a digital display. A screen appears to show a continuous image, but zoom in far enough and you reach the individual pixels, the minimum units below which the image cannot be further resolved. The Planck scale behaves in a similar way. It suggests that spacetime is not infinitely smooth and continuous but is composed of discrete units, minimum packets of space and time that cannot be subdivided further. A universe made of discrete units is, by definition, a digital universe. It is a universe that is quantized, divided into countable, finite pieces rather than existing as a seamless continuum.

Again, this does not prove simulation. Physicists offer other explanations for the Planck scale that do not require the universe to be computed. But the observation remains. Reality has a resolution limit. It is built from discrete units. And these are properties of designed systems, not properties that we would necessarily expect from a universe that exists independently and continuously.

Beyond the structural evidence from physics, there is another category of evidence that is harder to quantify but equally difficult to ignore. The glitches.

We introduced the Mandela Effect in the foreword. The phenomenon where large groups of people share detailed, consistent memories of events or facts that do not match the current state of reality. Nelson Mandela dying in prison. The spelling of the Berenstain Bears. The cornucopia in the Fruit of the Loom logo. The position of certain countries on the map. The exact wording of famous movie lines. In each case, thousands or millions of people independently recall the same incorrect version of reality, with the same specific details, despite having no connection to one another and no obvious reason to share the same false memory.

The standard explanation is that human memory is unreliable. And it is. Memory is reconstructive rather than reproductive. We do not record events like a camera. We rebuild them each time we recall them, and the rebuilding process is subject to errors, biases, and external influences. This is well established in cognitive psychology and neuroscience, and it accounts for many individual cases of false memory.

But the Mandela Effect is not about individual memory failure. It is about mass, coordinated memory failure with consistent details across unconnected populations. When one person misremembers the spelling of a children's book series, that is unreliable memory. When millions of people across different countries, different age groups, and different cultural backgrounds all misremember it in exactly the same way, the explanation becomes less satisfying. False memories should be random. They should vary from person to person, reflecting individual experiences and biases. The fact that they are consistent, that large groups of strangers remember the same wrong thing in the same wrong way, suggests that something other than simple memory error may be at work.

If our reality is a designed system, the Mandela Effect has a straightforward explanation. The system has been updated. Variables have been changed. Details have been altered after the fact, and the memories of some subjects retain traces of the previous version. Not all subjects, because the update was applied to their memories as well. But enough subjects to create a detectable pattern. Enough to leave fingerprints of the edit.

This is speculative, and it is important to acknowledge that. There is no controlled experiment that can test whether the Mandela Effect is caused by system updates to a simulated reality. But the phenomenon exists. The consistency of shared false memories across large populations is documented and reproducible. And the best conventional explanation, that millions of unconnected people spontaneously generate the same specific false memory through normal cognitive error, requires a degree of coincidence that is itself difficult to accept without question.

Then there are the experiences that individuals report but that science has never satisfactorily explained. Deja vu, the sudden, intense feeling that you have experienced a moment before, exactly as it is happening now. Precognitive dreams, where people report dreaming of events that subsequently occur with specific, verifiable details. Near death experiences, where people who are clinically dead report vivid, structured experiences that share common elements across cultures, religions, and historical periods. The sense of being watched, a feeling so universal that virtually every human being has experienced it, and one that has no agreed upon neurological explanation.

None of these are proof of anything. Each one can be explained, or at least partially explained, through conventional neuroscience and psychology. Deja vu is thought to be a misfiring of the brain's familiarity recognition system. Precognitive dreams may be the result of pattern recognition and confirmation bias. Near death experiences may be caused by oxygen deprivation or chemical changes in the dying brain. The sense of being watched may be an artifact of the brain's threat detection systems.

But there is an alternative framework that accounts for all of them simultaneously. If we are inside a designed system, deja vu could be the subjective experience of a repeated sequence, a moment in the simulation that has run before. Precognitive dreams could be the result of information leaking from the system's projected trajectory into the consciousness of the subject. Near death experiences could be what happens when a subject's consciousness partially disconnects from the simulation during a system boundary event. The sense of being watched could be exactly what it feels like. Being watched. By an observer on the other side of the experiment.

This framework is not scientific in the traditional sense. It cannot be tested, falsified, or verified through controlled experimentation. But neither can many of the conventional explanations for these phenomena. The difference is that the conventional explanations are accepted because they fit within an assumed framework of material reality, while the alternative is rejected because it does not. The question is whether that assumed framework is correct.

There is one more piece of evidence worth considering, and it comes not from physics or psychology but from mathematics. The universe is, at every level, governed by mathematical relationships. Not approximately mathematical. Not loosely mathematical. Precisely, exactly, unreasonably mathematical. The physicist Eugene Wigner wrote a famous paper about this, calling it the unreasonable effectiveness of mathematics in the natural sciences. The fact that abstract mathematical structures, invented by human minds with no reference to the physical world, turn out to describe the behavior of the physical world with perfect accuracy is, in Wigner's words, a gift that we neither understand nor deserve.

Why should the universe be mathematical? If reality is a brute physical fact, an independently existing thing that simply is, there is no obvious reason why it should be describable by elegant equations. Physical things do not need to be mathematically simple. They do not need to follow clean, symmetrical rules. They do not need to be computable.

But designed systems do. Software must be mathematically structured because it is built from mathematics. Simulations must follow computable rules because they are computed. A reality that is precisely, elegantly, unreasonably mathematical is a reality that behaves exactly the way a designed system would behave. Not proof. But not coincidence either.

The sum of all of this evidence does not prove that we are inside a simulation or an experiment. No single piece of evidence is conclusive, and every piece can be explained within conventional frameworks if you are willing to accept the necessary assumptions. But the convergence of so many independent lines of evidence, all pointing in the same direction, creates a picture that is increasingly difficult to dismiss entirely.

The universe has a resolution limit. Its fundamental equations contain error correcting codes. Reality behaves differently when observed than when it is not. The physical constants are fine tuned to values that permit life with essentially zero margin for error. The Mandela Effect produces consistent, shared false memories across unconnected populations. Human consciousness generates experiences that conventional science cannot fully explain but that fit neatly into the framework of a designed system.

And across all of recorded history, the species living inside this system has universally and independently concluded that something built it, something is watching, and something has rules for how the inhabitants should behave.

Perhaps the most remarkable thing about the evidence for a designed reality is not how strong it is. It is how long we have been looking at it without recognizing what it might mean. The clues have been in front of us for centuries, embedded in our physics, our psychology, our shared experiences, and our oldest beliefs. We have studied them, catalogued them, debated them, and filed them away into separate academic disciplines where they sit in isolation, each one interesting on its own but seemingly unrelated to the others.

But when you put them together, when you lay the physics next to the psychology next to the historical record next to the behavioral science, the picture that emerges is not random. It is coherent. It has a shape. And that shape looks remarkably like what you would expect to find if you were a subject inside a controlled experiment, looking for evidence of the experiment from the inside.

Calhoun's mice could not look for evidence. They could not analyze the structure of their enclosure or question the nature of their reality. They lived and died inside the experiment without ever knowing it existed. We are different. We can look. We can analyze. We can question.

And the more we look, the harder it becomes to dismiss what we find.

Chapter Thirteen

THE GODS OF THE CAGE

Imagine you are a mouse in Universe 25.

You have never known anything outside the enclosure. The walls of the cage are the boundaries of your universe. The food that appears in the dispensers is a constant of your reality, as reliable and unexplained as gravity. The temperature is always comfortable. The bedding is always there. You did not build this world. You did not choose to be in it. You were born into it, and everything about it simply is, without explanation, without origin, without cause.

But sometimes you see something. A shape beyond the walls. A shadow that moves in ways that nothing inside your world moves. A presence that is there and then gone, vast and incomprehensible, operating on a scale and in dimensions that your mouse brain cannot process. You cannot understand what it is. You cannot communicate with it. You can only sense, in whatever way a mouse senses anything, that something exists beyond the boundaries of everything you know, and that this something is watching.

If you had language, you would tell other mice about it. You would describe the vast, powerful presence that exists beyond the world. You would say that it created your world, because everything in your world appears to have been placed there deliberately. You would say that it provides for you, because food and water appear without any effort on your part. You would say that it watches you, because you have felt its gaze. And you would call it God, because you have no other word for something so far beyond your comprehension.

This is not a stretch of imagination. It is a description of exactly what would happen if a conscious, language using species were placed inside a controlled environment managed by an intelligence operating from outside that environment. The inhabitants would perceive the experimenter as a deity. They would attribute the conditions of their world to this deity's will. They would interpret the experimenter's actions, the provision of food, the maintenance of the environment, the occasional interventions, as divine acts. They would build systems of belief around this presence. They would create rules they believed the presence wanted them to follow. They would tell stories about its nature, its intentions, and its plans for them.

They would do exactly what every human civilization in history has done.

Every culture on Earth, without exception, has developed the concept of a higher power. A creator who made the world. A presence that exists beyond the visible. A force that watches, judges, provides, and sometimes destroys. This is not a regional phenomenon. It is not the product of a single tradition that spread through contact and influence. It emerged independently, on every continent, in every era, among populations that had no knowledge of one another's existence.

The Aboriginal Australians, isolated from the rest of humanity for at least 50,000 years, developed a complex spiritual framework centered on the Dreamtime, a creative period when powerful beings shaped the world and established the rules that govern it. The indigenous peoples of the Americas, separated from Eurasia for thousands of years, developed their own rich traditions of creator figures, spirits, and cosmic forces that built and maintained the world. African spiritual traditions, Polynesian mythology, ancient Chinese religion, the belief systems of pre contact Japan, the spiritual frameworks of every isolated human group ever studied, all of them arrived, independently, at the same fundamental conclusion. Something created this. Something is watching. Something has expectations for how we behave.

The probability that every human culture on Earth independently invented the same basic concept by coincidence is worth considering. Evolutionary psychologists offer explanations. The human brain is wired to detect agency, to assume that events are caused by intentional actors. This served a survival function. The ancestor who heard a rustling in the bushes and assumed it was a predator survived more often than the one who assumed it was the wind. Over time, this bias toward detecting agency became hardwired, and it produced a species that sees intention everywhere, including in natural phenomena that have no intentional cause. Rain becomes a gift from the gods. Earthquakes become divine punishment. The sun becomes a deity that must be honored to ensure its return each morning.

This explanation is reasonable. It may even be partially correct. But it does not fully account for the specificity and consistency of religious belief across cultures. The agency detection theory explains why humans might attribute natural events to intentional causes. It does not explain why virtually every culture converged on the same specific structure. A creator who built the world. An authority who established rules. A judgment that awaits those who break the rules.

An afterlife or continuation that follows physical death. A history of intervention, punishment, and renewal by the higher power.

These are not vague, generic spiritual impulses. They are specific, structural elements that repeat with remarkable consistency across cultures that had no contact with one another. And they are precisely the elements you would expect to find if the inhabitants of a controlled experiment had become aware, in whatever partial and distorted way, of the experiment and the experimenter.

Consider the Ten Commandments. In the Judeo Christian tradition, these are the foundational rules given by God to Moses on Mount Sinai. They are presented as divine law, the non negotiable behavioral code that the creator of the universe established for humanity. They carry the weight of absolute moral authority because they come from the highest possible source. God himself, speaking directly to his creation.

Now look at them from the perspective of experimental design.

Thou shalt not kill. Thou shalt not steal. Thou shalt not bear false witness. Thou shalt not covet thy neighbor's possessions. Thou shalt not commit adultery. These are not abstract moral principles. They are behavioral guidelines designed to prevent exactly the kind of social collapse that Calhoun observed in his mice. Killing destroys the population directly. Stealing disrupts resource distribution and creates conflict. False testimony destroys trust, which is the foundation of social cooperation. Coveting destabilizes the hierarchies that maintain social order. Adultery breaks pair bonds and disrupts the reproductive framework that sustains the population.

Every single one of these commandments addresses a behavior that, left unchecked, leads directly to the behavioral sink. Every single one of them is a rule that, if Calhoun had been able to communicate with his mice, he might have given them in an attempt to prevent the collapse he had already witnessed in previous experiments.

And that is the critical detail. Previous experiments. The Ten Commandments were not given at the beginning of the biblical narrative. They were given after a long history of failure. Adam and Eve broke the first rule they were given. Cain killed Abel. Humanity became so corrupt that God reportedly destroyed nearly all of it in a flood and started over with Noah. Then humanity failed again. And again. By the time Moses received the commandments, the story had already established a pattern of the creator attempting to guide human behavior, watching humans fail, punishing or resetting the population, and trying a different approach.

This is not theology. This is experimental protocol. A researcher who has run multiple iterations of an experiment and observed the same failure mode each time will eventually attempt to introduce behavioral constraints to see if the outcome changes. The Ten Commandments, viewed through this lens, are not divine moral law. They are revised protocol. The experimenter observed that the subjects consistently destroyed themselves through killing, stealing, deception, and social disruption, and introduced a set of explicit behavioral rules to see if constraining those specific behaviors would produce a different result.

The parallels extend far beyond the Ten Commandments. The concept of the flood exists in almost every major religious tradition on Earth. The biblical flood of Noah. The Sumerian flood in the Epic of Gilgamesh. The Hindu story of Manu and the great deluge. The Greek myth of Deucalion. The Aztec flood myth. The Aboriginal Australian flood narratives. Dozens of cultures, spread across the globe, with no historical contact with one another, all telling stories of a time when the creator destroyed the world with water and allowed only a small remnant to survive and rebuild.

The standard explanation is that ancient civilizations lived near rivers and coastlines, experienced real floods, and mythologized those experiences into creation narratives. This is plausible for some of the stories. But it does not explain why the structure of the myths is so consistent across unconnected cultures. It is not just that a flood happened. It is that the flood was sent by a higher power as a deliberate act of destruction in response to the failure of humanity. That a small group was chosen to survive. That the survivors were given a fresh start in a world wiped clean. That the flood was, in every version of the story, a reset.

If the history of life on Earth includes multiple extinction events and near extinction events, as the geological and historical record clearly shows, and if the inhabitants of the experiment have partial, distorted awareness of these events passed down through generations of oral tradition and mythologized over time, then flood myths are not fiction. They are memory. Garbled, symbolic, culturally filtered memory of actual resets. Events where the population was reduced to a remnant, the environment was transformed, and the experiment was allowed to continue with a fresh starting point.

The same logic applies to creation stories. Nearly every religious tradition begins with an account of the world being deliberately created. The details vary enormously, reflecting the cultural context of each tradition. But the core structure is the same. The world did not always exist. It was brought into being by a conscious, intentional act. An intelligence decided that the world should exist, and it made it so.

This is the universal human intuition that something built this place. And while science offers its own creation narrative in the Big Bang, the fundamental question remains unanswered. What caused it? What existed before it? Why did it produce a universe with physical constants precisely calibrated to support life? The scientific narrative describes the mechanism but not the intent. Religious narratives describe the intent but not the mechanism. If both are partial descriptions of the same event, the creation of a controlled environment by an intelligence operating from outside it, then both are telling the same story in different languages.

Now consider the most provocative element of the religion as experiment framework. The existence of multiple religions.

If there is one experimenter and one experiment, you would expect one consistent set of instructions. One religion. One rulebook. One clear communication from the creator to the creation. The existence of multiple, often contradictory religious traditions seems, at first, to undermine the hypothesis. If God is the experimenter, why would the message be inconsistent? Why would different populations receive different instructions?

There are two possible answers, and both are fascinating.

The first is that there are multiple observers. Different experimenters, or different research teams, running overlapping studies on the same population within the same environment. Each team communicates with its own subset of subjects, providing its own set of instructions calibrated to its own research questions. Christianity, Islam, Hinduism, Buddhism, Judaism, and every other major tradition represent different experimental protocols being applied to different groups within the same cage. The conflicts between religions, the wars, the persecutions, the centuries of hostility between groups who received different instructions, are the result of test groups interacting with one another in ways that the experimenters may or may not have intended.

This would explain why religions share so many structural elements, creation stories, moral codes, concepts of judgment and afterlife, while disagreeing on the specifics. They are different versions of the same basic communication, adapted by different research teams for different populations, all aimed at the same fundamental goal. Preventing the behavioral sink. Keeping the subjects from destroying themselves.

The second possibility is that there is one observer who has changed methodology over time. One experimenter, running a single long term study, who has tried different approaches to influencing the subjects' behavior across different populations and different eras.

The God of the Old Testament is strict, punishing, and interventionist. He destroys cities. He sends plagues. He floods the world. He gives rigid, absolute commandments and punishes those who break them with swift, often devastating consequences. This is an early stage experimental approach. Heavy handed control. Direct intervention. Immediate consequences for noncompliance.

The God of the New Testament, as presented through the teachings of Jesus, shifts dramatically. Compassion replaces punishment. Forgiveness replaces retribution. The message becomes about inner transformation rather than external compliance. Love your neighbor. Turn the other cheek. Judge not. This is a revised methodology. The strict approach did not produce the desired results, so the experimenter shifted to a gentler, more internally focused set of instructions.

Islam introduces a different framework. Disciplined structure, clear guidelines for daily behavior, a strong emphasis on community, submission to a higher purpose, and a detailed code of conduct covering everything from finance to family life to personal hygiene. This is another methodological variation. Not punishment based like the early Old Testament approach. Not purely internal like the New Testament approach. A middle path emphasizing structured behavior and communal accountability.

Buddhism takes yet another direction. Strip away the external rules entirely. Focus on the mind. Observe your own thoughts and impulses. Recognize the illusions that cause suffering. Detach from desire. Find peace not through compliance with external commandments but through internal awareness and discipline. This is perhaps the most sophisticated experimental approach of all. Instead of trying to constrain the subjects' behavior through rules, it attempts to modify their consciousness directly. To change not what they do but how they perceive and process reality.

Each of these approaches addresses the same fundamental problem. A population of conscious beings inside a closed environment, trending toward self destruction. Each one represents a different strategy for preventing or delaying that destruction. And none of them has worked permanently. Every major religion has produced periods of stability and periods of collapse. Every set of instructions has been followed for a time and then abandoned or corrupted. The experiment has continued through all of them, and the trajectory of the population has remained, over the long term, the same.

There is one more element of religious tradition that fits the experimental framework with disturbing precision. Prophecy. Almost every major religion includes predictions about the end of the world. Revelation in Christianity. The Day of Judgment in Islam. Ragnarok in Norse mythology. The Kali Yuga in Hinduism. The end of the current world age in Mayan cosmology. These are not vague, hopeful predictions about things getting better. They are specific, detailed descriptions of civilizational collapse, divine judgment, and the destruction or transformation of the existing world.

If the experimenter has run this experiment before, the experimenter knows how it ends. The experimenter has seen the behavioral sink play out in previous iterations. The experimenter knows the trajectory, the phases, the symptoms, and the outcome. Prophecy, in this framework, is not supernatural foreknowledge. It is projected outcomes based on data from previous runs. The experimenter communicated to certain subjects what was going to happen, not because the experimenter could see the future, but because the experimenter had already seen this particular future play out before. Multiple times. With the same result.

The religions promised that God would return. That there would be a final judgment. That the creator would come back to the experiment to evaluate the results, reward the compliant, and punish the failures. Every major tradition includes some version of this promise. The return of Christ. The coming of the Mahdi. The final avatar of Vishnu. The gods returning at the end of the age.

But what if the experimenter has decided not to come back this time?

What if, like Calhoun with Universe 25, the decision has been made to let this run play out without intervention? No more prophets. No more commandments. No more resets. No more floods. Just observation. Data collection. A final, uninterrupted run to see how the population behaves when left entirely to its own devices.

The silence of the modern age suddenly takes on a different meaning. The absence of new religious revelation, the decline of existing religious frameworks, the growing sense that nobody is watching and nobody is coming, these are not signs of human progress beyond the need for religion. They may be signs that the experimenter has stepped back from the cage for the last time.

HINDUISM
CHRISTIANITY
ISLAM
BUDDHISM

Chapter Fourteen

THE SILENCE

Calhoun ran twenty four experiments before Universe 25. In each one, he intervened. He adjusted variables. He introduced new mice. He modified the environment. He changed the starting conditions. He tried different configurations, different population sizes, different enclosure designs. Each experiment was an attempt to find a combination of factors that would prevent the behavioral sink, that would produce a population capable of sustaining itself indefinitely in a perfected environment.

None of them worked. The outcome was always the same. The population grew, thrived, peaked, deteriorated, and collapsed. The specific timeline varied. The specific details differed from run to run. But the trajectory was identical across all twenty four experiments. The behavioral sink was not a product of any particular configuration. It was a product of the fundamental conditions of the experiment itself. A closed environment with unlimited resources and no external challenges produced social collapse and extinction regardless of how the variables were arranged.

Universe 25 was different. Not in its design. Not in its outcome. In its methodology. Calhoun made a decision with Universe 25 that he had not made with any of the previous runs. He decided not to intervene. No adjustments. No introductions of new subjects. No modifications to the environment. He would set up the initial conditions, introduce the founding population, and then step back. Completely. Permanently. He would observe and record, but he would not interfere. Universe 25 would be the definitive test. The uncontaminated run. The experiment in its purest form, free from the variable of the experimenter's own involvement.

The result was total extinction. Every mouse died. The population curve rose, peaked, and descended to zero without a single intervention to slow or redirect it. Calhoun got his answer. When the experimenter stops interfering, the behavioral sink runs to completion. There is no self correction. There is no internal mechanism within the population that recognizes the decline and reverses it. The collapse, once begun, is irreversible.

Now consider the history we have discussed in this book through the lens of intervention and withdrawal.

The geological record shows five major extinction events over the last 450 million years, each one wiping out the majority of life on Earth and creating the conditions for new life to emerge and fill the vacuum. These were the great resets. Whether they were natural catastrophes or deliberate clearances of the cage depends on your interpretation, but their effect was the same. The experiment was restarted with new variables and new subjects each time.

The historical record shows multiple near extinction events within human history itself. Toba. The Bronze Age Collapse. The Black Death. The Plague of Justinian. Events that devastated the human population but did not eliminate it. Events that reduced the number of subjects dramatically but left enough survivors for the experiment to continue on a different trajectory. These were not full resets. They were interventions. Corrections. The equivalent of Calhoun adjusting a variable in one of his first twenty four experiments. Enough interference to change the conditions, not enough to end the run.

And the religious record, as we explored in the previous chapter, describes exactly this pattern from the perspective of the subjects inside the experiment. A creator who built the world, watched the subjects fail, destroyed most of them, allowed a remnant to survive, gave them new instructions, watched them fail again, intervened again, gave them revised instructions, and repeated this cycle through multiple iterations. The Bible alone contains a sequence of interventions that mirrors the structure of Calhoun's experimental series. Creation. Failure. Flood. Fresh start. Failure. Commandments. Failure. Prophets. Failure. Each time, the experimenter trying a different approach. Each time, the subjects eventually returning to the same destructive patterns.

This pattern of intervention is consistent across religious traditions, and it maps directly onto the pattern of near extinction events in the historical record. The stories describe what the data shows. Repeated attempts to redirect a population that keeps trending toward collapse. Repeated adjustments by an outside force that is trying, through various methods, to produce a different outcome.

And then the interventions stop.

When was the last time a new major religion emerged? When was the last time a figure appeared who was recognized across cultures and continents as a genuine prophet, a direct communicator with the divine, a carrier of new instructions from the creator? When was the last time the experimenter sent a message?

The major world religions all trace their origins to figures who lived centuries or millennia ago. Christianity to Jesus, roughly two thousand years ago. Islam to Muhammad, roughly fourteen hundred years ago. Buddhism to Siddhartha Gautama, roughly twenty five hundred years ago. Hinduism's foundational texts are even older. Judaism's core narrative stretches back thousands of years. Since the emergence of Islam in the seventh century, no new religious tradition has gained global traction on a comparable scale. There have been smaller movements, offshoots, reformations, and reinterpretations. But nothing that represents a genuinely new communication from the experimenter. No new set of instructions. No new prophet recognized across civilizations. No new revelation.

Fourteen hundred years of silence.

In human terms, that silence has been interpreted in various ways. Secularists see it as evidence that religion was always a human invention, and that as societies mature and science advances, the need for prophets and revelations naturally fades. Religious believers see it as a test of faith, a period where humanity must rely on the instructions already given and prove its ability to follow them without further guidance. Some traditions hold that the silence will end, that one final messenger or event will come to bring the experiment to its conclusion.

But there is another interpretation. One that fits the pattern we have been tracing through this entire book.

The silence is not a test. It is not a transition. It is a decision. The experimenter has stopped intervening.

Consider the sequence. For thousands of years, the record shows repeated interventions. Floods, plagues, commandments, prophets, new instructions, revised protocols. The experimenter was actively involved in the experiment, adjusting variables, communicating with subjects, attempting to steer the population away from collapse. Each intervention failed to produce a permanent change. Each new set of instructions was followed for a time and then abandoned or corrupted. Each reset led to a fresh start that eventually trended back toward the same outcome.

At some point, the experimenter made the same decision Calhoun made with Universe 25. Enough. No more adjustments. No more interventions. No more revised protocols. This run will proceed without interference to its natural conclusion, whatever that conclusion turns out to be.

The decline of religion in the modern world takes on a different significance in this context. The standard narrative is that humanity is outgrowing religion. That science, education, and rational thought are replacing superstition and faith. That the decline of religious belief and practice is a sign of progress, of a species maturing beyond its need for mythological frameworks and invisible authorities.

But what if the decline of religion is not humanity outgrowing the experimenter's instructions? What if it is the instructions being withdrawn? What if the fading of religious authority, the emptying of churches and temples, the growing sense among billions of people that nobody is listening when they pray, is not a triumph of reason but a consequence of abandonment? The experimenter is no longer reinforcing the message. The behavioral guidelines that were introduced to prevent collapse are being allowed to decay. The subjects are being left to their own devices for the first time in the history of the experiment.

And look at what is happening as a result.

Every symptom of the behavioral sink that we have discussed in this book has accelerated in the period since religious observance began its decline. Social structures are weakening. Birth rates are falling. Isolation is increasing. Mental health is deteriorating. Trust is eroding. Purpose is evaporating. The behavioral guidelines that once held societies together, whether you believe they came from God or from an experimenter or from the collective wisdom of the species itself, are being abandoned. And in their absence, the trajectory that Calhoun documented in his mice is asserting itself with gathering force.

This is not an argument for returning to religion. The instructions, whatever their origin, clearly did not solve the problem permanently. Thousands of years of religious practice did not prevent wars, did not eliminate greed, did not stop the strong from exploiting the weak. The commandments were given and broken. The prophets spoke and were ignored. The experimenter intervened repeatedly and the outcome never fundamentally changed. The behavioral sink was delayed, perhaps. Moderated, in some periods and some places. But never prevented.

The question is not whether the old instructions can save us. They have already been tried, across multiple traditions, across multiple centuries, and they did not produce a population capable of sustaining itself indefinitely. The question is what happens now that even those imperfect restraints are being removed.

Calhoun's answer, from Universe 25, is unambiguous. When the experimenter stops intervening, the collapse accelerates and runs to completion. There is no internal corrective mechanism within the population that can substitute for external guidance. The mice did not develop their own behavioral rules to replace the ones that the social structure had previously enforced. They did not form new communities or create new social frameworks to fill the vacuum left by the collapse of the old ones. They simply continued to deteriorate, faster and faster, until there was nothing left.

The modern world is conducting the same experiment in real time. The external frameworks that once constrained human behavior, religious, communal, institutional, are weakening simultaneously. Nothing of comparable force is replacing them. Individual freedom has expanded enormously, but freedom without structure is not liberation. It is disorientation. A mouse freed from all social constraints in Universe 25 did not flourish. It became a beautiful one. It withdrew, self isolated, and groomed itself into oblivion.

Humans freed from all social constraints are doing the same thing. Not all of them. Not yet. But the trajectory is visible, and it is accelerating. More isolation. More withdrawal. More self obsession. More purposelessness. More of every symptom that Calhoun documented as markers of a population in terminal decline.

The religious traditions promised that the silence would end. That God would return. That a final judgment would come. That the experimenter would step back into the experiment one last time to evaluate the results and determine the fate of the subjects.

Maybe that will happen. Maybe the silence is temporary, and an intervention is coming that will change the trajectory.

But Calhoun's Universe 25 offers a different possibility. One that is harder to hope around. In Universe 25, the experimenter stepped back and never returned. The observation continued. The data was collected. But no help came. No adjustment was made. No revised protocol was introduced to give the failing population a new chance.

The experimenter watched. Took notes. And let the cage empty itself.

The religions told us someone was watching. The physics suggests someone might be. The history shows a pattern of intervention followed by a growing silence. And the behavioral data shows a species that is tracking, with increasing precision, the same trajectory that has ended in extinction every time it has been observed.

The silence may be a pause. A test. A temporary withdrawal before a final revelation.

Or the silence may be exactly what it sounds like.

The experimenter has made a decision. This run will not be interrupted. This time, the data will be complete.

And we are the data.

CONCLUSION

The mice never understood. They lived their entire lives inside a world that had been built for them, maintained for them, and observed from the outside by an intelligence they could not perceive and could not comprehend. They were born into paradise. They thrived in it. They destroyed themselves within it. And at no point, from the first generation to the last, did a single mouse look at the walls of the enclosure and wonder why they were there.

We are different. We have looked at the walls. We have asked the questions. We have studied the structure of our environment with instruments and mathematics and centuries of accumulated knowledge, and we have found things that are not easy to explain. A universe fine tuned to impossible precision. A reality that behaves like a system being rendered on demand. Error correcting codes embedded in the equations that describe the fabric of spacetime. A planet calibrated to support life with a specificity that defies the odds. And a species that has independently, universally, across every culture and every continent, concluded that something built this place and something is watching.

We have also found a pattern. A behavioral trajectory that has been documented in controlled experiments, recorded in the geological and historical record, and measured in the demographic data of the modern world. A trajectory that begins with growth and prosperity and moves through predictable phases of social fragmentation, psychological deterioration, reproductive decline, and eventual collapse. A trajectory that has been observed in every closed system where it has been tested, and that has never, in any observed instance, reversed itself once it passed a certain threshold.

Calhoun saw it in his mice. Twenty five closed environments with unlimited resources and no external threats. Twenty five populations that grew, thrived, peaked, and destroyed themselves from the inside. He changed the variables. He adjusted the conditions. He tried different approaches across different runs. The outcome never changed. The behavioral sink was not an anomaly. It was a law. A predictable, repeatable response to a specific set of conditions. Give a population everything it needs, remove every external challenge, and watch it consume itself. Not sometimes. Every time.

The conditions that produced the behavioral sink in Universe 25 are now present on a global scale. Not in every country. Not in every community. But across the most developed, most prosperous, most technologically advanced societies on Earth, the same conditions that Calhoun created artificially in a metal box have been created organically by the march of human progress. Unlimited food. Eliminated predators. Controlled climate. Solved survival. The cage is bigger and more complex than anything Calhoun could have built, but the fundamental conditions are identical.

And the results are matching.

Birth rates across the developed world have fallen below replacement level and continue to drop. This is not a temporary fluctuation. It is a sustained, accelerating trend visible in every wealthy nation on the planet, resistant to every policy intervention that governments have attempted. The species that conquered every environmental challenge it has ever faced is quietly, steadily choosing not to replace itself.

Social structures that held communities together for centuries are dissolving. Family formation is declining. Marriage rates are falling. Community organizations are emptying. Religious congregations are shrinking. Civic participation is dropping. The web of relationships and obligations and shared commitments that once gave individuals a place within something larger than themselves is thinning to the point of transparency. What remains looks like a society from the outside. From the inside, it feels increasingly like a collection of strangers occupying the same space.

Mental health across the developed world is deteriorating at rates that have no historical precedent. Depression, anxiety, and suicide are rising fastest among young people in the wealthiest, safest, most comfortable nations on Earth. The generation with the most access to food, shelter, education, healthcare, entertainment, and communication technology in the history of the species is also the generation reporting the highest levels of psychological distress ever recorded. The paradise is real. The despair inside it is equally real.

The beautiful ones are everywhere. Disengaged, self absorbed, retreating into digital worlds that demand nothing and provide nothing of lasting value. Millions of people spending hours every day performing their lives for an audience of strangers while forming no meaningful bonds with anyone. The mouse in the corner grooming its fur has been scaled up to a global phenomenon, and it is so normalized that most people do not recognize it for what it is. It is not a lifestyle choice. It is a symptom. The same symptom Calhoun identified as the final stage before a population's decline becomes irreversible.

The aggression has lost its purpose. Wars are fought over resources that exist in abundance. Wealth is hoarded beyond any rational need while billions lack the basics. Political systems have fragmented into hostile factions that view each other as enemies rather than fellow participants in a shared society. The competitive drives that once pushed the species forward are misfiring in an environment that no longer provides meaningful outlets for them, producing conflict without function, hostility without cause, destruction without purpose.

And through all of it, the silence continues. No new prophets. No new revelations. No new instructions from whatever built this world and watched over it for so long. The frameworks that once constrained the worst of human behavior and gave communities a shared sense of meaning are fading. The churches are emptying. The traditions are weakening. The rules are being forgotten. Whatever was guiding the experiment, if anything ever was, appears to have stepped back and left the subjects to find their own way.

Calhoun stepped back from Universe 25 and watched the mice find their way. Their way led to extinction. Not through a dramatic catastrophe. Not through a single devastating event. Through a slow, quiet fade. Fewer births. More withdrawal. Less connection. Less purpose. Less reason to continue. The colony did not end with a bang. It ended with a whimper. The last mice dying alone in a paradise that had everything except a population willing to live in it.

This book has traced the parallels between that experiment and our own civilization. It has examined the evidence from physics, from history, from religion, from behavioral science, and from the demographic data of the modern world. It has not claimed to prove that we are inside an experiment. Proof may be impossible from inside the cage. But it has shown that the convergence of evidence, from so many independent fields, pointing in the same direction, is too consistent to dismiss as coincidence and too uncomfortable to accept without consequence.

If we are inside an experiment, then the resets that punctuate the history of life on Earth were not random catastrophes. They were previous runs that failed. The religions that emerged across every culture were not superstitions. They were the subjects' partial awareness of the experimenter. The commandments and moral codes were not arbitrary rules. They were behavioral protocols designed to prevent the collapse that had been observed in every previous iteration. And the silence of the modern era is not progress. It is the sound of an experimenter who has decided, after countless interventions that failed to change the outcome, to let this run complete without interference.

If we are not inside an experiment, the parallels remain. The behavioral sink is real whether or not an experimenter is watching. The data on social collapse, reproductive decline, and psychological deterioration does not require a simulation hypothesis to be alarming. The trajectory we are on leads to the same destination regardless of whether the cage was built by an intelligence or by the blind mechanics of physics and evolution. The pattern does not care about the origin of the conditions that produce it. It only cares that the conditions exist. And they do.

We have something the mice did not have. We have awareness. We can see the walls. We can read the data. We can study the pattern and recognize ourselves within it. We have the cognitive capacity to understand what is happening, to name it, to measure it, and to project where it leads. No previous set of subjects in this cage, if that is what it is, has ever had that ability. We are the first to see the experiment from the inside.

But awareness without action changes nothing. Recognizing a pattern is not the same as breaking it. Naming the behavioral sink does not reverse it. Publishing the data on falling birth rates does not raise them. Diagnosing the loneliness epidemic does not cure it. Identifying the beautiful ones does not bring them back to the world. We have been aware of these trends for years, in some cases for decades, and the trajectory has not changed. The curve continues downward. The warnings have been issued and ignored. The data has been collected and filed away. And the cage continues to operate exactly as it always has.

Calhoun spent the last years of his career warning anyone who would listen that what he had observed in his mice was beginning in human civilization. He pointed to the data. He drew the parallels. He issued the warnings. Most people did not listen. Not because the evidence was weak. Because the implication was unbearable. The idea that the most successful species on the planet, the species that split the atom and walked on the moon and mapped the human genome, could be subject to the same behavioral laws as mice in a metal box was too much for most people to accept.

It is still too much for most people to accept. We tell ourselves that we are different. That our intelligence, our technology, our self awareness make us immune to the forces that destroyed Calhoun's colonies. That we can think our way out of a biological trap. That progress will continue, that solutions will emerge, that the downward curve will reverse because we are too smart, too capable, too advanced to follow the same path as a colony of mice.

Maybe we are right. Maybe human consciousness really is different enough to break the pattern. Maybe the awareness we have, the ability to see the cage and understand the experiment, gives us a chance that no previous set of subjects has ever had. Maybe we will be the first population in the history of this experiment, if that is what it is, to recognize the behavioral sink and choose a different path.

Or maybe we are telling ourselves exactly what a population in decline always tells itself. That it is special. That the rules do not apply to it. That the pattern will not hold. That something will change before it is too late.

The mice could not tell themselves anything. They had no language, no abstract thought, no capacity for self reflection. They simply followed their biology into extinction without ever knowing what was happening to them.

We know what is happening. We can see the walls of the cage. We can read the results of the previous experiments. We can measure our own decline in real time, in spreadsheets and databases and peer reviewed studies and demographic projections that all point in the same direction.

We know. And so far, knowing has not been enough.

We thought progress was the answer. We thought comfort was the goal. We thought that solving survival would be the beginning of something greater. We thought that paradise was the destination, that once we arrived, everything would finally make sense, that we would find the purpose and meaning and fulfillment that every generation before us had been too busy surviving to achieve.

We thought we understood what we were building. Where we were going. What it all meant.

We got it all wrong.

Heinrich Wilson & Elias Verdan

HEINRICH WILSON
H.W.
PUBLISHING
BOOKS WHICH REFUSE TO BE SILENT

www.ingramcontent.com/pod-product-compliance
Lightning Source LLC
LaVergne TN
LVHW022051200926
848481LV00003B/871